# Contracting for Agricultural Extension
International Case Studies and Emerging Practices

# Contracting for Agricultural Extension
## International Case Studies and Emerging Practices

*Edited by*

## William M. Rivera

*College of Agricultural and Natural Resources*
*University of Maryland*
*College Park*
*Maryland*
*USA*

and

## Willem Zijp

*Operations Adviser, Africa Region*
*The World Bank*
*Washington DC*
*USA*

CABI *Publishing*

**CABI *Publishing* is a division of CAB *International***

CABI Publishing
CAB International
Wallingford
Oxon OX10 8DE
UK

Tel: +44 (0)1491 832111
Fax: +44 (0)1491 833508
E-mail: cabi@cabi.org
Web site: www.cabi-publishing.org

CABI Publishing
10 E 40th Street
Suite 3203
New York, NY 10016
USA

Tel: +1 212 481 7018
Fax: +1 212 686 7993
Email: cabi-nao@cabi.org



A catalogue record for this book is available from the British Library, London, UK.

**Library of Congress Cataloging-in-Publication Data**
Contracting for agricultural extension : international case studies and
Emerging practices / edited by W.M. Rivera and W. Zijp.
    p. cm.
Includes bibliographical references (p.    ).
   ISBN 0-85199-571-3
  1. Agricultural extension work--Contracting out--Case studies. I.
Rivera, William M. II. Zijp, Willem, 1948-
    S544 .C633 2002
    630'.71'5--dc21

                  2002003757

ISBN 0 85199 571 3

Printed and bound in the UK by Biddles Ltd, Guildford and King's Lynn, from copy supplied by the author.

# Contents

**Total off-loading**

# Contributors

## The Editors

**William M. (McLeod) Rivera** is Associate Professor in the College of Agricultural and Natural Resources at the University of Maryland, College Park. His experience includes service as consultant to international and national organizations, including field experience in more than 15 countries. Among his publications are three co-edited volumes: *Agricultural Extension Worldwide: Issues, Practices and Emerging Priorities* (1987), *Planning Adult Learning* (1987), and *Agricultural Extension: Worldwide Institutional Evolution and Forces for Change* (1991), as well as co-authorship with Willem Zijp and Gary Alex of the AKIS Good Practice Note on Contracting for Extension: Review of Emerging Practices.
Email: wr11@umail.umd.edu

**Willem Zijp**, Lead Specialist, Operational Quality and Knowledge Services, Africa Region, The World Bank, started his career in 1970 as an ox-trainer in Burkina Faso with the Dutch volunteer service. He subsequently worked in The Netherlands, Lesotho and Mauritania, for a total of 12 years of hands-on experience in Africa. In 1989 he joined the World Bank and worked on farmer empowerment, participation, sustainable agriculture and gender issues in extension. In 1997 he became – together with Derek Byerlee – one of the founding anchors of the Agricultural Knowledge and Information Systems (AKIS) thematic group in the Bank, and he was recently appointed to his present position in the Africa Region.

In addition, **Gary Alex** contributed to the editing of the first chapter. He is an independent consultant working with the World Bank on agricultural knowledge and innovation systems. He has 34 years experience in agricultural development, mainly in Asia and Latin America.
Emails: galex@worldbank.org; garyalex@erols.com

## The Contributors

The contributors are listed in order of regions represented (Africa, Asia and the Pacific, Europe, Latin America and the Caribbean, and North America) and the country on which their case study is produced.

### Africa

*Madagascar*

**Regina Laub-Fischer** is an agricultural economist and geographer who has been involved in advising in agricultural development, regional rural development and natural resource management since 1984. Since 1996 she has been working as technical advisor in the integrated forestry project implemented by the Malagasy Ministry of Water and Forest (MEF) and GTZ. Previously she worked in Madagascar as a short-term consultant for various development organizations. From 1984 to 1988 she provided technical assistance to a self-help project (GTZ) in Baluchistan/Pakistan.
Email: regina@dts.mg

*Mali*

**Eric Fermet Quinet** is a French livestock specialist (veterinary studies (PhD), master in animal production in tropical lands, training in epidemiology). He worked in many countries: Algeria, Mali, Senegal, Guinea Bissau, Yemen, Djibouti, Somalia, Palestine, Poland, Georgia, Chile and Peru. He is currently the head of the project PASPE (Support to the livestock private sector) in Mali.
Email: paspebamako@cefib.com

**Jérôme Gauthier** is a French livestock economist (veterinary studies (PhD), master in animal production in tropical lands, master in agricultural economics). He worked in many countries: Senegal, Ivory Coast, Mali, Burkina-Faso, Ghana, Chad, Cameroon, Uganda, Kenya, Zimbabwe, Botswana, Tanzania, Ethiopia, Tunisia, Yemen and Haiti. He currently works in the World Bank Division of Rural Development Washington.
Email: jgauthier@worldbank.org

*Mozambique*

**Helder Gêmo** is the Director of Agricultural Extension, Ministry of Agriculture and Rural Development, Maputo, Mozambique.

*Uganda*

**L. Van Crowder** has over 20 years of experience in teaching and training, research, extension and development communication. This includes working as Extension Communication Specialist and Associate Professor, Agricultural Education and Communication, University of Florida. He also served as Assistant Director, International Programmes, Institute of Food and Agricultural Sciences, University of Florida. International work has involved numerous technical assistance and advisory positions. Since 1993, L. Van Crowder has worked for FAO, first as Agricultural Education Officer and currently as Senior Officer, Communication for Development. He has a Masters in Communication and a PhD in Adult, Continuing and Extension Education, both from Cornell University.
Email: LoyVan.Crowder@fao.org

**Jon Anderson** is Environmental Advisor with the Asia and Near East Bureau of USAID. From 1994 to 2000 he was Forestry Extension Officer with the Food and Agriculture Organization of the UN in Rome. Prior to that he had worked on a World Bank funded forestry project in Mali and was project manager for USAID in Mali and Senegal. He has almost 20 years of rural development and natural resource management experience in Africa and has worked on extension issues in Mali, Mozambique and Tanzania as well as several countries in Asia. He has a Masters in Forestry from Yale University.
Email: janderson@afr-sd.org

**Asia and the Pacific**

*Australia*

Associate Professor **John Cary** is Principal Research Scientist in the Australian Bureau of Rural Sciences and a Principal Fellow at the University of Melbourne. He is a specialist in human behaviour and the management of natural resources, and has written extensively on modes of extension delivery. He has been a Visiting Fellow at the University of Michigan and the University of Wisconsin-Madison. He is co-author of *Greening a Brown Land: the Australian Search for Sustainable Land Use.*

**Andrew Campbell** is the Executive Director of the Australian Land and Water Resources Research and Development Corporation. Previously he was responsible for the development and implementation of Bushcare, funded through the \$1.25 billion Natural Heritage Trust. He was Australia's first National Landcare Facilitator from 1989 to 1992. He is the author of *Landcare—Communities Shaping the Land and the Future* and *Planning for Sustainable Farming.*

**Trevor Webb** is employed as a social scientist in the Social Sciences Centre of the Bureau of Rural Sciences. He has recently completed his PhD exploring the meanings of multiple use landscapes in the Great Barrier Reef World Heritage Area, and has a research interest in the human conferral of meaning to rural and agricultural landscapes.

*Bangladesh*

**M. Hassanullah** has 35 years' work experience in agricultural research, education and extension at home and abroad. Presently, he is working as Senior Specialist in the Agrobased Industries and Technology Development Project (ATDP), a project of the Ministry of Agriculture, funded by USAID and implemented by the International Fertilizer Development Center (IFDC). His recent works relate to development of agribusiness, crop diversification, strategic agricultural research and development planning. Dr Hassanullah is a life member of the Association for International Agricultural and Extension Education, Ohio, USA, the Indian Society of Extension Education and a former President of the Bangladesh Agricultural Extension Society.

*China*

**Yonggong Liu** is professor and vice dean of the College of Rural Development (CORD) and deputy director of the Center for Integrated Agricultural Development (CIAD) of China Agricultural University. After postgraduate studies at the Department of Animal Science of Beijing Agricultural University in 1985, he received advance training in the field of rural extension and rural sociology in the University of Hohenheim, Germany. Since 1991, besides teaching work at the university, he has been working as senior consultant and researcher for World Bank, ADB, FAO, UNDP and other international development agencies. His main research is focusing on participatory extension training methodology, rural development planning and resource management. Email: liuyg@public.bta.net.cn

*Vietnam*

**Dinh Duc Thuan** was educated in former East Germany, and after returning to Vietnam was appointed lecturer in Forest Economics at the Vietnam Forestry University (1979-1985). He then became Head of the Forest Economics Department (1986-1992), Following this, he was Director of the Forest Development and Social Forestry Centre (1993-1995). Since 1995 he has been

Director of the Social Forestry Training Centre, a semi-autonomous institution belonging to the University, the leading forestry education organization in Vietnam.

**Nguyen Ba Ngai** is currently Vice Director of the Social Forestry Training Centre at the Vietnam Forestry University. Since 1992 he has been a lecturer in Social Forestry and Forestry Extension, and has provided field training in participatory methods and extension. Mr Ngai completed his undergraduate studies in former East Germany and was a lecturer in Forest Economics prior to 1992.

**Bardolf Paul** works for Helvetas, a Swiss NGO and is Chief Technical Adviser for the Social Forestry Support Programme, a Swiss-funded initiative supporting the reform of tertiary level forestry education in Vietnam. Working in Vietnam since 1991, Mr Paul has been involved in introducing participatory approaches first in forestry extension and more recently in forestry education and training. He has been active in promoting contractual initiatives in extension training between the University and development projects.
Email: sfsp.bp@hn.vnn.vn

## Latin America and the Caribbean

*Chile*

**Julio A. Berdegué** was born in 1957 in Mexico, and has been a resident of Chile since 1984. An agronomist by training, he has studied in the former National School of Agriculture in Mexico; the University of Arizona at Tucson, the University of California at Davis, both in the USA; and the Wageningen Agricultural University in The Netherlands. He has led and participated in numerous agricultural research, extension and rural development programmes, with emphasis on the application of systems perspectives and methodologies to the analysis of small-scale agriculture in Latin America. He has been actively involved in developing Internet-based knowledge and information systems to support rural and agricultural development projects in poor rural communities. Mr Berdegué has co-edited four books on different aspects of systems-oriented research in agriculture and rural development, and is also the author of over 50 articles and book chapters.
Email: berdegue@reuna.cl

**Cristián Marchant** was born in Chile in 1954. He holds a BSc degree in Agronomy from the University of Chile. For 22 years he has worked at INDAP (Chile's Agricultural Development Institute), where he has served as Chief of Area, Regional Director, and national Chief of Operations.
Email: cmarchan@indap.cl

*Colombia*

**Rodolfo Rodríguez** holds the degree of MSc in Tropical Agricultural Development from the University of Reading (England). He is Executive Director of the Fundación CICADEP, the International Center for Agricultural Development and Training in Santafé de Bogotá, Colombia.
Emails: cicadep@atenea.lasalle.edu.co; rodolfo121@hotmail.com

**Tito Rodríguez** holds a BSc in Animal Production from the National University of Colombia. He works in the Projects Unit of the Fundación CICADEP in Santafé de Bogotá, Colombia.

*Trinidad and Tobago*

**Joseph Seepersad** PhD is Senior Lecturer, Faculty of Agriculture and Natural Sciences, University of the West Indies, Trinidad and Tobago. He has taught and conducted research in Agricultural Extension and Extension Communications for several years.
Email: seeps@tstt.net.tt

**Wayne Ganpat** MSc is Agricultural Officer in the Extension Division of the Ministry of Agriculture in Trinidad. He was directly involved in the development of the mass media strategies to combat the Pink Hibiscus Mealy Bug (HMB) threat.

**Europe**

*Estonia*

**Ülar Loolaid** graduated from Tartu University in 1981 in Mathematics, and pursued postgraduate studies in human geography. He is employed at the Institute for Rural Development Ltd as Leader for theme development. He is a member of the Estonian Association of Rural Consultants and the Commission for Extension Matters at the Ministry of Agriculture. His experience in extension includes applied research in monitoring and evaluation, developing participatory rapid appraisal methodology, and training advisory skills.
Email: mai@server.ee

*Finland*

**Riikka Rajalahti** currently works as a Project Manager for Widagri Consultants Ltd., a consultancy company working on Natural Resource Management, Gender, and Household Economics. Prior to joining Widagri, she was an Interim Professor and a Senior Research Associate at the University of Helsinki, Finland, and concentrated on research and teaching on Farming Systems, Weed

Science and Sustainable Natural Resource Management. She has a PhD from Cornell University, majoring in Vegetable Crops.
Email: Riikka.Rajalahti@widagri.pp.fi

**Eija Pehu** is President of Widagri Consultants Ltd and has served as the professor of crop production in the University of Helsinki for the past 10 years. Her research and teaching interests range from farming systems to crop physiology and applied agro-biotechnology. Professor Pehu has carried out over 30 consultancy missions to Africa and Asia on tropical cropping systems including support to developing agricultural extension services and rural financing. She is also a member of the Board of Trustees of the International Potato Center and of the Advisory Committee of the EU-Center of Technical Assistance for the ACP countries.
Email: Eija.Pehu@widagri.pp.fi

*Germany*

**Jochen Currle** is a member of the international consultancy group PACTeam, working on organizational and methodological questions of agricultural extension. After his studies of agricultural sciences and a research project in Peru on the possibilities of agricultural extension to protect natural resources, he worked as a Research Fellow at Hohenheim University. His PhD thesis focused on extension approaches and participatory technology development in the prevention of soil erosion.
Email: PACTeam@T-Online.de

**Volker Hoffmann** is Professor at the Department of Agricultural Communication and Extension, Institute for Social Sciences of the Agricultural Sector, Hohenheim University. He is also currently the President of the German-based network for research in the agricultural sector of the Tropics and Sub-Tropics, ATSAF, and a member of the Board of Trustees for the International Institute for Tropical Agriculture (IITA), Ibadan, part of the CGIAR system.
Email: i430a@Uni-Hohenheim.de

**Andrew D. Kidd** is a member of the international consultancy group PAC Team and a Visiting Fellow at the Department of Agricultural Communication and Extension, Institute for Social Sciences of the Agricultural Sector, Hohenheim University. He is concerned mainly with issues of institutional arrangements for service delivery in social and natural resource development, including agricultural extension.
Email: Kidd@ Uni-Hohenheim.de

*The Netherlands*

**Paul Duijsings** is senior expert in management of extension and advisory services. In 1980, following his study at the Agricultural University Wageningen, he began his career in the advisory service of the Ministry of Agriculture and Fisheries. In 1990 he became one of the directors of the new advisory service in The Netherlands, the DLV, and was responsible for the executive management of the division for extension and advice for farmers in livestock production (300 advisors). In 1998 and 1999 he became Chief Operating Officer of the entire advisory company. At the end of 1999 he became director of corporate development of the DLV Adviesgroep Inc., which includes business development, marketing, R&D and knowledge-management. More information about DLV at: http://www.dlv.nl

**Jet Proost** has a background in rural sociology and extension science. For several years she worked in Western Africa and in The Netherlands in project development and training of agricultural advisors. In 1991 she joined the Department of Communication and Innovation Studies at Wageningen University as assistant professor in environmental issues in agriculture and management of extension services. Her special interest is in farmer-driven change processes and the ways to support these processes. Besides her position at the university she works as a journalist and a private consultant for communication projects in agriculture. More information about Wageningen University and the department at: http://www.sls.wau.nl/cis
Email: Jet.Proost@Alg.VLK.WAU.NL

*Portugal*

**Artur Cristóvão** is Professor of Extension and Rural Development at the Department of Economics and Sociology, University of Trás-os-Montes e Alto Douro, Vila Real, Portugal.
Email: Acristov@utad.pt

**Timothy Koehnen** is Associate Professor of Extension and Rural Development at the Department of Economics and Sociology, University of Trás-os-Montes e Alto Douro, Vila Real, Portugal.

**Fernando Alves** is an agronomist, Director of the Association for the Development of Viticulture in the Douro Valley and responsible for the IPM programme described in the Portugal case study.

**North America**

*USA-Illinois*

**Burton E. Swanson** is Professor of International Agriculture, Department of Agricultural and Consumer Economics, University of Illinois at Urbana-Champaign. He received his PhD in Development Studies, University of Wisconsin in 1974. He has written or edited four books, plus numerous monographs, chapters, journal articles, and technical reports. He has participated in more than 75 assignments in 35 developing countries, primarily to help design, implement and/or supervise projects for international organizations.

**Mohamed Samy** is a Visiting Scholar, Department of Agricultural and Consumer Economics, University of Illinois at Urbana-Champaign. He received his PhD in International Agricultural Extension from the University of Illinois in 1988. He is an Associate Professor at Menoufia University in Egypt, and previously served as an Associate Professor, College of Agriculture, United Arab Emirates University. He currently serves as lead analyst for a value-added research project that seeks to improve Illinois farm income.

**Patrick D. O'Rourke** is Professor of Agribusiness and Agricultural Economics, Illinois State University. He received his PhD degree in Agricultural Economics from Purdue University in 1979. He was an Assistant Professor at Purdue University from 1979 to 1983 and was then appointed as an Assistant Professor at Illinois State University in 1983. He teaches nine subjects in agribusiness management, agricultural economics, foundations of inquiry, and international development at Illinois State University.

*USA-Louisiana*

**John Barnett**, Extension Cotton Specialist, Louisiana State University Agricultural Center, worked previously as a county agent conducting education programmes for row crop producers in Louisiana and Arkansas. Over a 25-year career in vocational agricultural and extension education, he developed agricultural curricula, published educational materials and technical articles, served as an officer in several professional associations, and is recognized for his expertise in cotton production and educational contribution.

**Satish Verma**, Programme and Staff Development Specialist and Professor of Extension and International Education, Louisiana State University has been involved in designing and evaluating extension education and training programmes in the Louisiana Cooperative Extension Service. He also carried out dairy extension work in India, and had institution-building assignments in Sierra Leone, Jamaica, and Ukraine focused on extension and outreach. Email: sverma@agctr.lsu.edu

# Foreword and Acknowledgements

This compilation of case studies on contracting for agricultural extension services results from an initiative by the editors to identify a wide variety of cases germane to the subject, discover what impact they have had to date, ascertain the likelihood of sustainability and replicability and determine what lessons had been learned. The Agricultural Knowledge and Information Systems (AKIS) Thematic Group of the World Bank, in collaboration with the University of Maryland, College Park then decided to compile a volume on contracting for extension services. The present compilation is the result of that original initiative and collaboration.

The editors contacted various specialists worldwide who were knowledgeable about the current developments shaping agricultural extension as well as experienced in the policy and operations of particular developments in their country involving contracting for extension services. The editors received over 30 case studies, kept 26 for analysis, but limited the selection to 18 examples for the purposes of this volume, which focuses specifically on contractual arrangements aimed at procuring agricultural extension services.

The term "extension" is used broadly herein and includes all agricultural information transfer programmes and activities by public, private and related sectors, and is intended to cover related terms, such as information transfer, technology transfer and advisory services. The term "outsourcing" and "contracting out" are used interchangeably to indicate arrangements whereby public sector agricultural and rural extension systems contract with private sector entities for services. The volume presents the growing body of experience worldwide with the contracting approach and its usefulness for improving the financing and delivery of agricultural knowledge.

## Context and Generalities

This volume presents the cases that served as the basis for the findings and discussion in the AKIS Good Practice Note issued in February 2001 (but dated December 2000), titled *Contracting for Extension: Review of Emerging Practices*. What that text and this volume suggest is that contracting for extension services – albeit not without disadvantages and not prescribed as a panacea for deep-seated system failures – is none the less a positive development and a vital strategy for the advancement of knowledge transfer in the agricultural domain.

The cases in this volume highlight a number of generalities, namely that contracting for extension is widespread, a strategy employed by all sectors, and utilized in a variety of extension situations. An analysis of the cases uncovers a number of facts regarding the technical criteria, the social and environmental consequences and the impacts of contracting for extension.

The findings on technical criteria relate to the selection, monitoring and evaluation and certification of advisory consultants, payment/cost sharing of the consultant service fee, the funding of programmes providing contractual advisory services, who decides the content of extension messages, and who decides who will receive the advisory service. The consequences for societies and the environment indicate effects regarding stakeholder participation, poverty, equity, food security, natural resources management, capacity building and other concerns, such as women and children; judging from the cases these effects tend to be positive. Additionally a number of considerations emerge: (i) the importance of political will to promote system reorganization and contracting for extension; (ii) the institutional roles, opportunities and benefits of contracting for extension; and (iii) the impact of contracting for extension on capacity building, changed roles, and the utilization of providers.


## Acknowledgements

The editors are grateful to numerous colleagues. We thank Gary Alex (World Bank) for his editorial assistance in preparing the AKIS Good Practice Note, *Contracting for Extension: Review of Emerging Practices* and for his invaluable comments on the organization of the present volume. We also extend our gratitude to the many colleagues both within and outside the World Bank who provided useful comments on the Good Practice Note, the comments having also contributed to our development of the present volume. William Rivera is especially appreciative of the assistance by his former student, consultant and friend, Jon Kurst, for his insights and help with the original analysis of the case studies.

We thank the AKIS thematic team for its support, in particular the chairs, Marie-Helene Collion and Derek Byerlee.

We are obliged especially to the contributors to the volume. These include: Trevor Webb, John Cary, and Andrew Campbell (Australia); Artur Cristovão, Fernando Alves, and Timothy Koehnen (Portugal); Rodolfo Rodríguez and Tito Rodríguez (Colombia); L. Van Crowder and Jon Anderson (Uganda); Dinh Duc Thuan, Nguyen Ba Ngai, and Bardolf Paul (Vietnam); Yonggong Liu (China); Julio Berdegué and Cristían Marchant (Chile); Ülar Loolaid (Estonia); Jochen Currle, Volker Hoffmann, and Andrew D. Kidd (Germany); Regina Laub-Fischer (Madagascar); E. Fermet Quinet and J. Gauthier (Mali); Jet Proost and Paul Duijsings (The Netherlands); Joseph Seepersad and Wayne Ganpat (Trinidad & Tobago); Burton E. Swanson, Mohamed Samy, and Patrick D. O'Rourke (United States/Illinois); John Barnett and Satish Verma (United States/Louisiana); M. Hassanullah (Bangladesh); and Riikka Rajalahti and Eija Pehu (Finland).

Not all of the submissions to the volume proved acceptable, and in some cases other duties demanded the attention of potential contributors before they could complete their country case studies. To all those who demonstrated interest and willingness to assist with the volume, we extend our sincere thanks.

# Introduction

William M. Rivera and Willem Zijp

In this era of unprecedented change, the institution and systems of agricultural extension continue to be reinvented (Rivera, 1996). Fiscal system reform, decentralization, privatization and democratization are the keywords and fundamentals of extension's contemporary change. The underlying pressures for reform are associated with globalization, the new economic system that has evolved from liberalization of trade, structural adjustment, technological and telecommunications advances, and greater interdependence worldwide.

At the same time, agricultural information has been changing in terms of its content, the means by which it is transferred, and its marketability as a "commodity". Its content has been changing since the chemical industry's entrance into the agricultural domain in the mid-1800s, and more radically since the Green Revolution of the 1960s. The means of transfer have been advanced, chiefly in high-income countries by the modernization of telecommunications and the popularization of computers, providing immediate access – for instance to information on farm commodity prices worldwide and localized weather conditions. As Zijp (1994) notes, information technology is making both public and private sector agricultural information systems more accessible and more rapid in transmission. The commodification of agricultural information, i.e. the transforming of knowledge into a product for sale, has begun to revolutionize both public sector extension and the business of private sector technology transfer.

The transformation of public and private sector agricultural information systems also results from changes wrought by larger exogenous and endogenous global forces. These forces include developments related to international trade and global competition, population dynamics, science and technology development, land use and the natural environment, structural changes in institutional development, the supply of and demand for trained workers, and poverty, illiteracy and poor quality of life. The challenge to respond to these many forces is rendered especially difficult because of limited public sector financial resources. Although many of the global forces just cited are exogenous to the individual country's agriculture and natural resource sector, they are central to any diagnostic of agricultural support systems and constitute the configuration against which to formulate and evaluate agricultural policies and programmes for these systems.

The world's ideology has changed. The "power shift" (Mathews, 1997) from public sector services to private sector hegemony, and the inexorable drive of individual companies to increase their control of production and market share reflect the new global ideology. This current ideology transition towards global capitalism and "free market" principles is expressed in aphorisms such as "the new world order" and "liberalized trade". The transition to this new ideology has been taking place from 1985 to the present and can be described as a period of downsizing public universities and research and extension services, as well as privatizing parastatals, and encouraging overseas private investments and new forms of public/private partnerships (Eicher, 1999).

Along with the new ideology, structural adjustment programmes have been imposed to bring developing countries into line with the financial demands of globalization. These structural adjustment programmes have strongly impacted the national governments of less developed countries. Many of these countries have been, and continue to be, pressured to reform their public sector systems. Public sector agricultural extension is, in increasing instances, pressed to adopt various decentralization, cost-recovery and privatization strategies (Rivera and Gustafson 1991; Smith, 1997). One notable development in this era of change has been that of public sector contracting for the delivery of extension services.

## Contracting for Delivery of Extension Services

Contracting for delivery of agricultural extension services is a widespread strategy used in many countries and situations. In developing countries, contracting often shifts delivery of extension from public to private providers. In some cases, this is reversed, with private sector entities contracting with the public sector for specialized services. Analysis of case studies from 18 countries on five continents underscores the variability in approaches to contracting for extension services.

Continued economic liberalization is likely to result in a growing number and greater diversity in extension service providers, as demand for new products, information, and services develops and incomes rise. Farmers and rural dwellers already have access to an increasing number of information sources. Steady improvement in rural infrastructure and rising standards of literacy will change the nature of demands, and continued government fiscal restraint will force reduction of subsidized state extension for market-oriented producers. These developments are likely to lead to a rise in the numbers and types of contracting and partnership arrangements for extension.

Public sector contracting for extension services can provide significant advantages in flexibility and increases in efficiency and effectiveness. Major policy issues need to be addressed: removing barriers that restrict development of private service providers; developing transparent contracting procedures; introducing cost awareness and cost analysis in programmes; addressing equity concerns with private service providers; and managing additional costs to the public sector in contracting extension. Although there are possible disadvantages as well as advantages to outsourcing extension services, this volume of case studies highlights good practices in some 18 countries worldwide.

## Organization of the Volume

The volume is organized into six main sections. The main text is followed by a multi-language bibliography including both textual and web site references, and an index to assist readers in their examination of the volume. Readers will note that some of the cases might have been placed in sections other than that in which they appear. For instance, the Portugal case deals with contracting IPM extension and just as easily might have gone into Section Two, Contracting to Promote Environmental Services. However, the editors decided that the case fitted more appropriately in Section Five, Farmers Contracting for Commercial Advisory Services. Also, many of the cases involve "off-loading" by the public sector, but they indicate other directions as well and were placed in sections highlighting other aspects of contracting for agricultural extension services. Some cases could fit in other, perhaps several sections. Advice to the reader: read all the cases to determine their possible connections within the framework of this volume on contracting as an emerging trend in extension reform.

## The Sections

The cases in section one concern off-loading public sector extension services to the private sector. These cases cover instances of totally shifting both funding and delivery of field extension services to a private venture company (The Netherlands), totally shifting only the responsibility for extension delivery to private consultant firms (Chile), and regularly shifting a variety of specialized tasks to private extension entities (Estonia). Cases in other sections of the volume also cite instances of shifting or sharing responsibility for extension delivery services (e.g. Germany, Mozambique) but these cases have been placed in later sections because of some other distinctive aspect of their contracting arrangement. Additionally, although not included in this volume, there are countries (e.g. Peru) where funding and delivery of services were shifted entirely to non-governmental organizations. Some countries (e.g. Venezuela) contract with multiple providers for the single purpose of carrying out extension field services. In short, this volume underlines much of the scope and variety of contracting for extension, but it is not an exhaustive treatment of all the variations on extension contracting arrangements. (Relatedly, in Chapter 1 see Table 1.1 Different arrangements for financing and implementing contract extension.)

Section one highlights the fact that the public sector has in a number of cases shifted partial or total responsibility to the private sector for provision of extension field services. In *Chile*, the government continues to fund, but does not provide a public sector service to deliver agricultural extension for small farmers. Certified private consultants provide extension information on demand from farmers. The *Estonia* case emphasizes the multi-faceted role of contracting for private sector agricultural advisory services. Three regionally different approaches to agricultural extension have developed in the *Federal Republic of Germany*. The *Thuringia* case underscores the move in the Eastern Region toward private sector provision of field services. A completely privatized advisory service has evolved in *The Netherlands*. Originally the public sector shifted its technical staff to farmer organizations, but later this staff was transferred to a private company, the DLV Adviesgroep Inc. Like Chile, *The Netherlands* has a system that has developed in response to changing conditions and requirements.

The next three sections contain cases involving specific contractual arrangements. Section two provides cases where contracting has been employed to promote environmental services. In *Australia* the state governments have fostered contracting with community organizations to prevent land degradation. Known as "Landcare" this initiative is considered a notable success story. *Madagascar* also presents a case where government has contracted with local communities to manage their own natural resources in a sustainable way. This contractual arrangement developed as a result of the Malagasy government's 1997 forestry law and the concept of "secured local management of natural resources".

Section three draws on the experiences of three countries in contracting for input services. The *Bangladesh* case presents the results of sub-contracting extension services to a local private agricultural training institute to train farmers in hybrid maize cultivation. *Mali* provides an example of contracting for livestock production extension with private veterinarians. In the case of *USA-Illinois*, farmers contract with high-tech companies for precision agricultural services.

Section four consists of three cases that involve contracting for specialized services. The *Colombia* case examines a semi-private coffee-growers contracting system. In *Trinidad and Tobago* the government contracts for extension communications services to initiate a Hibiscus Mealy Bug Information Campaign. The *Vietnam* case involves contracting for specialized extension training in participatory planning.

The fifth section changes pace, citing cases where farmers contract for commercial advisory services. In *Portugal's* Douro Valley, the Association for

Viticultural Development contracts for IPM extension. The case of *USA-Louisiana* is about private crop consulting services in the cotton industry of Louisiana and the implications of this contracting scheme for enhancing interdependency among private consultants, extension and farmers.

The final section highlights cases that demonstrate characteristics or tendencies distinct from those cited in earlier sections. The *China* case observes the development in Xinyang Prefecture in Henan Province, from government-driven group contracting extension to farmers' need-oriented contracting extension. *Finland* provides two examples of contracting, noting the country's long history of contracting for agricultural extension. *Mozambique* differs in that the contracting process is in it infancy, but it indicates the intention towards contracting for extension in one of Africa's largest and poorest countries. The plan is to contract initially with non-governmental organizations, while maintaining public sector extension services in remote areas. *Uganda* is an example of private sector secondment of government extension agents, a form of "contracting in" by the private sector to obtain the services of public sector technicians.

All in all, the cases indicate a wide variety of contracting arrangements, and suggest that contracting as a strategy is likely to continue to be an important means of procuring and delivering extension services – whether by the public sector, farmers, or the private sector.


## Availability on the Web

The present compilation is conceived as a work-in-progress. The volume, far from complete, nevertheless provides a relatively broad aggregation of countries and approaches to contracting for extension. In order to encourage access to the material and findings in this volume, the Good Practice Note and a number of related case studies are intended to be placed on the Web at *www.workbank.org/akis* (under AKIS Series: Good Practice Notes). Thus the present volume is intended to serve both as a stimulus to discussion and as a preliminary installment on the subject of contracting for extension developments worldwide.


## References

Eicher, C.K. (1999) Institutions and the African farmer. *Issues in Agriculture* 14. CGIAR, Washington DC.

Mathews, J.T. (1997) Power shift. *Foreign Affairs* 76(1), 50-66.

Rivera, W.M. (1996) Reinventing agricultural extension: fiscal systems reform, decentralization and privatization. *Journal of International Agricultural and Extension Educatio* 3(1), 63-67.

Rivera, W.M. and Gustafson, D.J. (eds) (1991) *Agricultural Extension: Worldwide Institutional Evolution and Forces for Change*. Elsevier, Amsterdam
Smith, L.D. (1997) *Decentralisation and Rural Development*. FAO/SARD, Rome.
The World Bank (2000) Contracting for extension: review of emerging practices. *AKIS Good Practice Note*. www.worldbank.org/akis (AKIS Series: Good Practice Notes).
Zijp, W. (1994) *Improving the Transfer and Use of Agricultural Information: a Guide to Information Technology*. The World Bank, Washington, DC, Discussion Paper.

# Chapter 1

# Good Practices in Contracting for Extension

William M. Rivera, Willem Zijp and Gary Alex

This chapter reproduces much of the AKIS Good Practice Note (2000). The general findings of the cases are presented along with a review of the policy issues relevant to contracting for extension services and the requirements suggested in the case studies for successful extension contracting.

In general, successful contracting of extension services requires: (i) political will for extension programme reform and contracted extension; (ii) clarity in institutional roles, opportunities, and benefits of contracted extension; (iii) adequate capacity of service providers; and (iv) effective demand for extension services. Contracted extension services require collaboration and partnerships characterized by good will, and all parties need to be results oriented. Clients (producers) will be more involved in selecting extension agents, evaluating extension services, certifying agents, sharing costs of programmes, contracting services, determining content of extension programmes, and deciding how services are allocated.

Numerous recommendations emerge from the case studies. These include the need for: (i) situation-specific analysis of objectives, and an assessment of the advantages and disadvantages of contracting-out of extension services; (ii) a change in the role of government from service delivery to facilitation and quality control; (iii) detailed terms of reference and contract documents that ensure clarity in both process and responsibilities; (iv) increased client ownership of programmes and more public-private partnerships; and (v) clarity in objectives and strategies when initiating a programme of contracted extension. Experience with contracted extension is still relatively new, and there is much to be learned from evaluation of these cases as they mature.

## General Findings

The cases tend to recommend contracting for delivery of extension for two reasons. The first is that contracting strategies tend to promote a greater number and variety of providers of agricultural extension information, and thereby encourage competition in an area that has been criticized for its ineffectiveness and inefficiencies. The second reason is that contracting strategies foster cost sharing by clients, which helps to increase responsiveness and accountability of service providers to clients.

There is a growing acceptance that state monopoly of agricultural extension service delivery is neither desirable nor sustainable. Agricultural extension programmes are under pressure to change, because of growing fiscal pressures and questions about effectiveness and efficiency of their service. Changing views of agriculture and the role of government, new opportunities in market economies, and pressures for increased public participation and good governance are leading to a reassessment of the cost-effectiveness and the relevance of agricultural extension (Farrington, 1994). Strategies increasingly being employed with private service providers in extension programmes are cost recovery, commercialization, privatization, and subcontracting (Rivera and Cary, 1997).

Extension reforms are most advanced in industrial countries where previously state-owned and managed extension services have been completely or partially privatized (for example, New Zealand, United Kingdom, The Netherlands). The impact of these reforms has yet to be fully evaluated, but there appears little, if any, move to return to previous systems of organization and governance. This suggests that the clients of extension – producers and other stakeholders – are not dissatisfied with the transformation and welcome the evolution to more truly demand-driven services.

## Contracting for Delivery of Extension Services

The need to reform public extension services is perhaps greatest in developing countries, where initiatives for cost recovery, commercialization, and contracting service provision are becoming more commonplace. Major reforms of state extension services, some including radical innovations, have been implemented (Chile, Mexico, Colombia, Nicaragua) or are underway (Zimbabwe, Zambia, Uganda). Tools being used to improve extension efficiency include decentralization of government services, increased pluralism of service providers, and development of more client-responsive extension methodologies to increase customer "ownership" and influence on extension services. Contracting for services and cost sharing with private sector for-profit firms and individuals, non-governmental organizations (NGOs), and farmer organizations are on the increase.

Current views on the optimum role of government stress the need to decentralize and devolve; to downsize, cut costs, and become more efficient; and to develop partnerships with other actors (Rivera, 1996). In parallel with general trends in civil service reform, there is an increasingly common view that, while government may need to continue to fund extension services, the actual delivery of these services can, in many cases, be subcontracted to private sector providers. In this volume, "private sector provider" refers to all "non-state sector providers", including private for-profit firms, NGOs, individuals, farmer and community organizations, civil society organizations, and private educational institutions.

Conventional wisdom holds that government extension is ineffective and inefficient and has been too monolithic, heavy handed, and controlling. There is concern that governments have created extension bureaucracies that are over-staffed, have little funding for operating expenses, use unsustainable approaches, and are overly supply-driven, particularly in centralized economies.

Issues regarding extension reform are well known and discussed in-depth elsewhere (Smith, 1997; Carney, 1998; Feder *et al.*, 1999), but one frequent response to these perceived problems is that of contracting out (or out-sourcing) extension delivery to the private sector (either for-profit or not-for-profit). For example, by hiring more efficient NGOs to provide extension services, governments can trim public payrolls and use savings for operating costs that make remaining staff more operational and effective.

## Separating Functions of Financing and Service Delivery

Contracting is a means of providing incentives for individual extensionists and service agencies (state and nonstate) to deliver specified outputs (farmers trained, demonstrations presented, information delivered, advisory services provided). Outputs must be specified in contracts, which provide for payment only on completion of outputs, and which may provide special incentives for efficiency or effectiveness in completion of outputs. Contracts can also provide incentives based on achievement of results or outcomes of the contracted services (increased producer incomes, productivity, or even poverty reduction). Contracting along these lines between township extension bureaus and producers is widely practiced in China.

Contracting for extension services sounds quite straightforward, but review of current experience reveals a wide range of approaches to separating functions of financing, procurement, and delivery of services. Alternatives include both public and private financing, public and private contracting for services, and public and private service delivery. Options also vary with the type of public or private financier, contracting entity, or service provider. The public sector includes national, state or provincial, or local government. The private sector encompasses for-profit business and consulting firms, nonprofit and civil society institutions, individual farmers and farmer organizations.

## Options for Contract Financing and Service Delivery

The best known arrangements for funding and delivery of agricultural information services are *private-financing/private-delivery* (i.e. contract farming) and *public-financing/public-delivery*. Contract farming, an independent issue from that of public sector reform, is not discussed in this volume. The current reform of public sector extension services results from economic and political attacks on their efficiency and effectiveness. As a result of these criticisms and the various other pressures for reform enumerated in the volume's Introduction, the public sector has devised various new arrangements for funding and delivering extension services. Public sector involvement in extension has rightly continued in part because purely private extension systems generally lack incentives to address public concerns, such as long-term social and environmental issues.

Two other options mix public and private sectors in financing and service delivery, and are the basis for most new contracting initiatives. *Public-financing/private-delivery* (i.e. "contracting out" or outsourcing) is the approach commonly promoted in reform of extension services. A different option, that of *private-financing/public-delivery* has been neglected generally, but field experience indicates that it is prevalent in many countries. It may have significant potential to provide farmers and farmer organizations with quality services. In some countries, such as Uganda and Mozambique, NGOs hire well-known public extension advisors to help provide services. In other countries, such as Israel, farmer organizations contract with public sector extension for specialized services.

*Contracting Private Provision of Public Services*

In contracting out extension delivery, public funds are used to contract private providers of services. Anticipated benefits include greater operational efficiency and cost-effectiveness; greater accountability of extensionists to perform and produce results; and a greater variety of providers of extension services. When publicly financed extension services are contracted out, the role of government changes from that of implementing agency to that of quality controller, overseer, and provider of training and technical information to agencies contracted. In some incipient arrangements, as in Mozambique, both the public and private sector may provide services, with division of labour by district and province.

With private sector service delivery, extension workers can be rewarded more easily for good performance and dismissed for poor performance. Although there are obvious and well-documented problems with public sector extension systems, there is no guarantee that extension provision by the private sector is going to be more effective. White and Eicher (1999) observed that "there is a lack of conclusive evidence that NGOs are more efficient than government or private agencies in delivering farmer support services".

Chile, Hungary, and Venezuela achieved successful public-funding/private-delivery of extension services. In Chile, services tend to be provided by private for-profit firms, in Hungary by universities, and in Venezuela by NGOs. Such cases illustrate the range of potential service providers and the options for contracting services from universities, NGOs, or farmer organizations when contracting with for-profit private service providers is not an immediate possibility.

*Commercializing Publicly Provided Services*

Although emphasis is generally on public sector contracting for services, there are frequent instances of cost-recovery or fee-for-service systems in which private sector entities contract the public sector for delivery of specialized services. In these situations the private sector pays for, and the public sector delivers, extension services. Fee-based public extension services exist throughout Europe. Commercialized public extension recognizes that many extension services are private goods for which users are willing and able to pay, but for which economies of scale and scope make it more efficient for government to develop and deliver the service.

NGOs, farmers, farmer organizations, private companies, religious organizations, and other groups often enter into agreements with government extension agents (usually at the local level) to provide services (particularly technical advice) to their members or clients. These entities are generally agreeable to agents staying with the government not only for economic reasons, but also to ensure linkages with the government and to facilitate access to public services and resources. Moreover, NGOs and private companies often find cost advantages in collaborating with government extension services, as they can tap technical skills on an as-needed basis without carrying staff on their payrolls.

## Case Studies of Extension Contracting Arrangements

This review of 18 cases of extension contracting confirms the extensive use of contracting as a means of enhancing the design, funding, delivery, and monitoring of extension. The cases display a wide variety of sources of financing for extension contractual arrangements, and an equally wide variety of extension service providers operating under contract (see Table 1.1). Although useful, the matrix in Table 1.1 tends to oversimplify the reality of field extension services. Funding and delivery of extension services often involve mixed public and private arrangements with joint financing and collaboration between the public and private sectors in service provision.

Contractual arrangements for extension fall within two main groupings: two-party contracts and multiparty contracts.

**Table 1.1.** Different arrangements for financing and implementing contract extension.

| Source of financing | Provider under contract | Case study examples |
| --- | --- | --- |
| International donors | Public institution | No example in this volume |
| | Private for-profit organization | Mali |
| National governments | Farmer organization | Australia, Finland, Portugal |
| | Private NGO | Vietnam |
| | Public sector institution | No example in this volume |
| | Private for-profit organization | Bangladesh, Chile, Estonia, Trinidad |
| State or local governments | Public sector institution | China |
| Universities | Public sector institution | Finland |
| Farmers or farmer organizations | Private for-profit organization | USA/Louisiana, USA/Illinois, The Netherlands |
| Multiple sources | NGO | Colombia, Mozambique, Uganda |
| | Public sector institution | No example in this volume |
| | Private for-profit organization | Germany, Madagascar |

*Two-party contracts*

The review found four types of two-party contracts, each with several variations:

**1.** Donor projects provide financing for government agencies to subcontract for extension services. The Bangladesh case presents contracting arrangements for procuring extension services from private sector experts.
**2.** Governments contract services through ministries of agriculture, regional governments, and national institutes. In Trinidad and Tobago the government extension unit contracts with mass media to provide extension messages. And,

in Vietnam, the Ministry of Agriculture contracts training services from the National Forestry University.

**3.** Farmer organizations contract public or private sector entities for extension advice. In Portugal the Association for Viticultural Development in the Douro Valley contracts for IPM extension services. In the USA-Louisiana case farmers pay for private row cotton crop consulting.

**4.** Individual farmers contract directly with private entities for extension advice. Chile provides public funding for small farmers to contract private delivery of agricultural extension. Germany provides public funds for farmers to contract with registered extension providers.

*Multi-party Contract Arrangements*

When several partners are involved in extension service delivery, costs for all must be considered in evaluating system efficiency. These "coalition systems", however, seem to have advantages that go beyond economic/staffing considerations, and may relate to functional complementarities, institutional comparative advantage, and commitment to effective service provision. The many arrangements involving several partners seem to reflect attempts to draw on complementarities and strengths of different partners rather than simply a search for adequate financing.

Multiple partner agreements may promote sustainability by being both demand-driven and responsive to public concerns, such as soil and water conservation, food security, and gender equity. The three types of multiparty contracts found in the cases studies involved varied agreements between producers, service suppliers, and funding agents. The multiparty contracts were between:

**1.** A local municipality, producers, and an investment fund (Colombia).
**2.** Local populations, service organizations, and a ministry (Madagascar).
**3.** Farmers, extension service centres, and other partners (China).

Although the contractual arrangements found in the study do not represent an exhaustive list of possible arrangements, they underscore the variety in contracting arrangements. These may include contracts between: official government and private entities; universities and private organizations; projects and private associations; individual farmers or farmer organizations and institutes, cooperatives, and projects; and agribusinesses and farmers. Contracts may be for commodity-specific, specialized or general extension services.

## Policy Issues in Contracting Extension Services

The efficiency and effectiveness of contracted extension services depend on many of the same factors that influence any extension programme.

Macroeconomic and fiscal instability increases cost of capital and reduces the propensity for private investment. Inadequate protection for property rights reduces producers' ability to secure credit and invest in long-term productive enterprises. Government controls and market interventions reduce the efficiency of input supply and output markets, and thus limit farmer ability to respond to market incentives. All of these policy issues condition the ability of extension – whether via contract or not – to impact on productivity and innovation in agriculture.

Still, strategies for contracting extension services – particularly public financing of private service delivery – present some important policy issues for national extension systems. Many countries have few private sector service providers. Contracting procedures are important, but may vary considerably. Relative cost-effectiveness and cost-efficiency of public and private providers is important. Social and equity issues must be considered, as must programme impacts on government budgets and state capacity to manage contracted extension services.

## Reducing Barriers to Entry for Private Sector Providers

The limited existence and availability of competent nonstate providers in many developing countries reduces choice and competition, thereby restricting opportunities for contracting extension services. This may be due to the scarcity of private organizations with a mission to serve farmers through agricultural advisory services, lack of demand for information services, excessive government regulation, or government services having "crowded out" private participation. Expansion of private sector capacity could improve producer access to a range of services and meet the needs of a diverse clientele.

## Developing Explicit Contracting Procedures

The cases indicate that public contracting of private service provision can be politically sensitive, if there is even the appearance of nepotism or unfair competition in the award of contracts. Contracting processes require complete transparency through the use of clear procedures and accountability mechanisms, such as certification of service providers, competitive contracting processes with broad-based selection committees, and regular reporting and routine financial audits. Accordingly, competitive selection procedures are desirable to ensure fairness in contractor selection, and to establish competitive prices for contracted services.

Contracting for extension demands procedures that are clearly defined and documented, with:

1. Legal contract documents to formalize agreements between parties.

**2.** A detailed "Operation Manual" describing contracting procedures and roles and responsibilities of parties involved.

**3.** Clear terms of reference to guide the bidding process, outlining the type of proposals expected, and objectives of the contract. Services must be specified sufficiently to establish a contract for: work to be done, responsibilities of the parties, geographic area (if applicable), target clients, and products or outputs expected.

**4.** A contracting process that advertises, widely for service providers, and provides adequate time and information for potential providers to prepare and submit proposals.

**5.** A technically qualified selection committee to evaluate proposals.

**6.** Contracts signed promptly upon award.

**7.** A procedure to inform all parties of the selection results, giving unsuccessful bidders information on how to improve proposals for subsequent calls for proposals.

## Comparative Cost-efficiency and Impact Effectiveness

Cost/benefit estimates and cost accounting are important aspects of extension programme management, but few extension staff seems adequately trained to carry out these tasks. Contracting procedures that require attention to unit costs create strong pressures to keep costs in line, and to purchase services from suppliers having competitive cost advantages (Scott, 1996).

Private providers are expected to use resources more effectively or be more likely to deliver desired outputs and goals, since most public extension services appear to neglect cost-effectiveness indicators and lack real incentives for good staff performance. Costs are likely to vary from country to country, and will depend on the type and intensity of services. As a reference point, extension advisory service programmes in Nicaragua and Mozambique are estimated to have annual unit costs per farm family of about $400.

## Social and Equity Considerations

The case studies underscore the varied impacts contracting extension services have had on social and environmental issues. In ten cases contracted services addressed poverty issues and increased farmer incomes. Three cases reported greater equity in the distribution of benefits from extension, but four reported a negative impact on poor farmers. Contracting positively affected women farmers and labourers in five cases, and stakeholder participation in programme decision making increased in 14 cases. Seven cases had positive environmental outcomes and contracting enhanced capacity of institutions to provide advisory services in 14 of the 18 cases.

Social and equity issues are important considerations for extension policy, especially for countries with a high proportion of small, resource-poor farmers and endemic rural poverty. Public policy generally seeks to direct services to poverty alleviation and social equity, while for-profit private firms are thought to be less concerned with these issues than is the state sector. (However it must be admitted that government agencies' record of achievement with poverty and social programmes leaves much to be desired.) Gender-equity also requires that contracts consider women's issues in service delivery.

How can contracted extension systems avoid having a negative impact on services for the poor? Some cases suggest that contracting extension services can specifically target small, poor, and marginal farmers, and require contractors to work with these groups and address poverty issues. Private sector for-profit providers are willing to service difficult areas under favourable contract terms, and NGOs often have greater aptitude for working effectively with disadvantaged communities. Government policy needs to ensure that the poor have access to services needed to improve social indicators and productivity. Contracts can also target the poor by stratifying clients according to income level, as is done in some Latin American extension programmes (Chile, Mexico).

## Additional Costs When Outsourcing Extension Services

The decision to contract provision of extension service entails long-term planning and costs. Managing contracted extension services requires skilled staff with the capacity to monitor, supervise, and evaluate the work of contractors. Financial management and administration systems in many government agencies also require strengthening to handle contracted extension. Other costs are the loss of institutional memory, when contractors develop exclusive expertise and retain experience with extension programmes. In the extreme, this can lead to an effective monopoly on service provision. A further cost is the loss of key government staff, who might be presented with comparatively attractive employment conditions in the private sector.

Outsourcing services might also require heavy initial costs, if reforms include staff redundancies and retirement packages in addition to financing of contract services. Political and social concern over downsizing public sector staff is frequently an obstacle to contracting services from private providers. However, a well-managed programme of training, grants, and severance packages to assist staff into retirement and new occupations can mitigate this problem. Over the medium term, private sector expansion leads to a progressive increase in private sector employment opportunities and facilitates rationalization of government staff numbers and cost. Indeed, the development of new services and innovations by the private sector may result in more people being employed in extension, as in New Zealand, where extension has been totally privatized and there are now more extension consultants than when extension was a state monopoly (Walker, 1993).

# Requirements for Successful Extension Contracting

This volume focuses mainly on government outsourcing of extension services for farmers, though many of the principles are equally valid for other contracting systems. While the cases tend to emphasize field advisory services, some cover other activities, such as group formation, mobilization, marketing, product preparation, and farm management. In all cases, the requirements for successful contracting of extension were similar.

## Essential Preconditions for Publicly Contracted Extension

A sound policy environment for agricultural development (including farmer access to inputs and markets) is basic to any meaningful extension service. The case studies reveal four additional preconditions for successful programmes of contracted extension: political will to promote system reform, clarity in institutional roles, adequate capacity in service providers, and effective demand for extension services.

### Political Will to Promote Extension Reform

Reform of agricultural extension is an urgent need in many countries. If government operations are improved by contracting for service delivery, farmers can receive better services, public expenditures can be reduced, and private enterprise can be expanded. Change threatens all, however, especially vested interests, and the shift to contracted extension implies a fundamental change in the traditional view of government.

For contracted extension services to be successful, adequate funding is required – whether from donors and government, the private sector, farmer organizations, or a combined effort by the various parties concerned. There must be a willingness to cooperate among all parties, including politicians, service providers, and clients. Government agencies must be responsive to clients' needs, whether expressed directly or though the contracted service provider. Building trust and political will for reform is the essential first step in initiating a strategy of contracting extension services.

### Clarity in Institutional Roles

The government role needs to be spelled out and clearly understood in systems of contracting services. Governments have an essential role in looking after public goods and the public interest. This frequently requires government to maintain a role in training extension advisers, contract oversight, programme monitoring, programme evaluation, and overall strategy formulation. Contracting for extension ideally involves co-production of services with collaboration between government, service providers, and farmers.

Contracting arrangements provide the opportunity for extension providers (including government, NGOs, private venture companies, and farmer organizations) to share experiences and resources, and to engage in collaborative planning. Collaborative planning with clients encourages a sense of local ownership and tends to enhance programme effectiveness. Contracts that ensure client participation in decisions on content and delivery of services strengthen demand-driven systems, and increase the relevance of information services provided.

*Adequate Capacity in Service Providers*

One of the common constraints to contracting extension services is the real or perceived lack of qualified service providers. Barriers to entry cited above limit the number of institutions engaged in provision of extension services, but often institutions are active in the field, even though "invisible" to government officials and agencies. Establishment of a registry of service providers is helpful to facilitate the contracting process. Private sector agencies, both for-profit and non-profit, are innovative and quick to enter a field, when funding is available.

Contract extension arrangements can exploit the comparative advantage of service providers. However, provider professionalism and technical capacity are fundamental to success. Training and education (through strong coalitions among agricultural education institutions, research and extension) may be essential in the short term for upgrading, and in the long term for advancing a professional consulting industry.

*Effective Demand for Extension Services*

The other aspect of the technical services market is the demand-side. This is usually weaker than the supply-side of service providers, as farmers are usually small and not formally organized. Furthermore, past experience with top-down extension services has not stimulated a great appreciation for the value of technical services. Public programmes to finance or co-finance services for farmers, and approaches that put the farmer in charge of the extension agenda, tend to strengthen demand for extension services.

Farmer organizations are critical to strengthening the demand for technical services, because they provide economies of scale and a mechanism for promoting small-farmer interests. Technical competence of farmer organizations is important in dealing with service providers and farmer education and training are crucial to enhance agricultural development over the long term.

**Operational Requirements for Contracting Extension**

Contractual arrangements often involve an evolutionary process and move through phases before reaching maximum efficiency and equity (as noted in the cases on Chile and The Netherlands). Success depends on finding practical solutions to local problems. The case studies reveal seven operational issues important to contracting for extension.

*Selection of Extension Agents*

Success of extension activities depends largely on extension agents and their relationship with farmers. Selection of extension agents is therefore critical to programme success, and contracting services should allow greater discretion and flexibility in selecting them. *Farmers or members of farmer organizations should play a greater role in selecting the extension agents who serve them.*

*Monitoring and Evaluating Contract Extension Services*

Government, sometimes with input from NGOs and farmers, is generally responsible for monitoring and evaluating the work of contracted extension agents. Each extension contract should include a plan to monitor the performance of the consultant and/or the consultant firm. Evaluation records should be kept to maintain a record of available consultant firms for extension work, and to record the assessment of the firm's performance. Capacity to monitor and evaluate contracts is crucial.

Time-bound milestones should be included in contracts, and contract objectives should be clear and backed by verifiable monitoring indicators. In Zambia a separate unit has been established dedicated to this purpose, and a similar development is planned in Zimbabwe (Ashworth, 2000). Monitoring and evaluation mechanisms must provide for monitoring both performance of contract service providers and impact of contracted services. *Farmers and farmer organizations should be more involved in monitoring and evaluating the work of extension agents.*

*Certification of Extension Agents*

A system of professional accreditation is needed for quality control and monitoring qualifications of extension service providers and/or extension agents. In most cases, government is responsible for maintaining a registry and certifying capacity of private service providers financed by public funds. The certification process can be problematic and can open government staff to charges of abuse or favouritism. Farmers, professional associations, and other private entities also maintain professional registries, evaluate consultant capabilities, and validate credentials, so farmers should probably be more involved in evaluating extension agents and certifying their capabilities. *Private sector or civil society mechanisms for certifying extension agents and service providers should be a long-term objective in contract extension systems.*

## Payment/Cost Sharing of Extension Costs

Government continues to provide funding for most contractual extension arrangements, but cost-sharing arrangements between governments and farmers are increasingly prevalent. In industrial countries, farmers are paying either a large portion or all costs for extension. Farmers also co-finance extension services from the for-profit private sector, NGOs, and research institutions. *Cost-sharing arrangements with farmers and other agencies should be encouraged in publicly financed contracted extension programmes.* Cost sharing is greatly facilitated when farmers are involved in selecting, monitoring, and evaluating extension agents, and in determining programme content.

## Contracting Extension Services

As governments move to outsource extension service provision, the initial approach is usually that of government contracting services on behalf of farmers. Although this is expedient and government maintains a certain level of comfort with the management of public funds, it maintains a top-down mentality and a relationship of government doing something for the client. Empowering farmers to purchase their own services gives them true responsibility for service provision. Many community and producer organizations are weak and poorly organized, but experience indicates that they frequently can contract services efficiently and effectively (De Silva, 2000). *Programmes should rely less on government contracting and encourage farmers and farmer organizations to assume responsibility for contracting extension services.* This is likely to be a gradual process in most countries.

## Deciding on the Content of Extension Messages

Governments continue to determine the extension messages to be provided to farmers, although they now more often do so in consultation with NGOs. Farmers and the for-profit private sector are less involved in determining content of extension services. *Farmers, NGOs, the private sector, and research institutions should be closely involved in decisions on extension programme content.* Diversity in sources of technology (research, private sector) is important, especially in countries where extension must serve many types of farmer.

## Deciding Who Will Receive Extension Services

The decision as to who should receive public services is very much a political decision and will be affected by the political process. To the extent that clients value services and lobby for them, this is a legitimate means of allocating services. There are, however, important public goods issues involved in targeting services. Equity concerns dictate that services be targeted to the poor, women, or minority groups, to address specific problems. Environmental concerns may dictate a need to focus activities on a particular area or problem. A broad range of stakeholders has interests in, and need to be consulted on, allocation of extension services. *Governments should consult widely with stakeholders and use objective indicators in decisions on who will receive extension advisory services.*

# Recommendations for Practitioners

An analysis of the cases suggests a number of recommendations for best practice in contracting for extension. Originally intended to assist policymakers in planning contractual arrangements for extension services, the editors hope that the following recommendations will also stimulate practitioners to further thinking and discussion of this emerging practice.

### Strategies for Contracting Extension Services: Diagnostic vs. Recipe

Experience indicates that design of programmes for contracting extension varies greatly from country to country. While some general lessons may serve to guide practitioners, local diagnostic work is essential and "blueprint" approaches are to be avoided. Case studies illustrate the fact that contracting for extension can have varied objectives and various contractual arrangements. Governments may wish to get out of the business of extension delivery and shift that responsibility to private organizations. Alternatively, a public extension agency may simply need to expand its capacity by contracting extension providers to cover certain specific areas. Programme design also depends on the socioeconomic and agricultural systems prevalent in the area to be served. An agency might contract for services in one area and not necessarily in others (livestock but not crops, or forest seed production but not development of fishponds).

Contractual arrangements may involve two, three, or more parties, and variation in contract arrangements can lead to confusion. Systems are best characterized on the basis of the entities carrying out three major functions – financing, purchase of services, and provision of services. Each of the three functions can be split between public (local, regional, or national level government) and private sectors (NGOs, farmers, agribusiness, or for-profit firms). Increasingly, demand-driven extension systems involve central government providing financing for local government or farmer groups to contract services from private sector providers.

### Redefining the Role of the State

With contractual extension arrangements, the public sector role changes from service delivery to ensuring quality of the services provided under contract. The days of government command and control over extension delivery are numbered, but the role in oversight and regulatory functions continues to be very important. The quality control function extends beyond contract oversight to addressing second generation problems, especially the links between service providers and support services that ensure quality extension – subject matter specialists expertise, research and development, training, and monitoring and evaluation.

Governments will likely continue to *influence* extension processes so that public goods are taken into consideration. As Christoplos and Nitsch (1996) state: "Hopes that the private sector and civil society will simply expand to fill the vacuum in basic services have proven to be enormously exaggerated." Public agencies must have the capacity to design sound terms of reference, set the rules of the game for fair tendering processes, monitor contract implementation, and evaluate the results.

More experience is needed to define good practice in facilitating the transition from traditional to contractual extension systems. In particular, there is an issue of what to do with a public service system existing in parallel with the contracted services. Contractual extension arrangements are not appropriate to all situations, although they are likely to spread to new areas where they have potential to address problems facing many traditional extension systems. International organizations, such as The World Bank and the FAO/UN, need to be able to assist their member countries to assess and design appropriate programmes for contracting extension services.

## Initiating Programmes to Contract for Extension

When initiating a programme of contracting for extension, situation-specific assessments of objectives, advantages and disadvantages of contracting extension services are necessary. Pilot projects could prove useful to establish procedures and best practice and to encourage development of private service providers and appropriate partnerships. On the negative side, contracting extension services can result in services being provided only to those able to pay or in areas most easily served. Identifying the target client population in well-designed terms of reference is important. Contracting with NGOs with a special mandate or methodology for reaching the poor should help ensure equity.

## Ensuring Clarity in Contract Documents and Process

Detailed terms of reference and good contract documents and procedures are important to ensure properly functioning programmes. Terms of reference should be specific about deliverables and the timetable for delivery, as well as expected results to be achieved. The full contracting process includes preparation of requests for proposals, legal contract documents, and a procedural manual outlining the precise roles and responsibilities of all parties. When government is issuing the contract, responsibilities for oversight and regulation must be spelled out, along with procedures at all levels (central, provincial and local) for monitoring and evaluating performance of the contracted service provider.

**Fostering Public-Private Partnerships**

Many countries/areas lack significant numbers of potential service providers, though local NGOs and other organizations are emerging to fill this role. The question remains as to how best to promote the emergence of private sector service providers. Clearly, government must provide a policy environment that fosters emergence of private services. As farmers and the private sector develop greater capacities, and as demands on agricultural systems expand, new partnerships and institutional relationships will prove essential to link farmers, government, and private institutions. Many of these relationships will be based on sharing information and co-production of services. Complementarities between institutions are important, and partnership arrangements can exploit those complementarities.

**Expanding Beneficiary Ownership of Extension Services**

The traditional manner of financing extension services involves a top-down flow of funds from the government or donor to the service provider. With very few exceptions, the chain of accountability for performance is upwards to the financier and producers; service users are excluded from the chain. Reversing the financial flow by providing funds directly to the producer to contract services from providers can change the entire incentive and accountability structure. This fundamental reversal of funding and accountability has several advantages in that:

1. Providers must deliver relevant services of value to their paying customers.
2. Services are more likely to have a positive cost-benefit ratio since producers are unlikely to pay for services they do not consider beneficial.
3. Services are more likely to be sustainable after the withdrawal of state/donor funding.
4. Empowering beneficiaries promotes competition among providers and spurs further efficiency gains.
5. Introducing cost-sharing arrangements is facilitated, because producers are already directly involved in financing and will purchase services they truly need and value.
6. Poor and marginalized individuals and groups can be provided with real buying power.

**Monitoring Future Experience with Contracted Extension**

Much is yet to be learned about systems for contracting extension. These systems are generally too new to show much evidence of impact of moving from traditional to contracted services. The case studies were generally positive and/or claim good impact, but none of the case studies had any sound impact evaluation. There is therefore no evidence of reduced costs under contracting systems, although it seems likely that effectiveness has increased. Bank support would be useful to develop evidence of impact and cost-effectiveness. Issues worth further study include: empirical analysis of true unit and aggregate costs of contracting extension, mechanisms for fostering ownership, continuity of contracted extension systems, and the potential to enhance sustainability of contractual extension arrangements by linking them with other services (input supply, credit, marketing).

# References

Ashworth, V. (2000) *Contracting for Extension: Some Principles and Emerging Practices for Consideration.* Analytical paper produced for AKIS. The World Bank, Washington, DC.

Carney, D. (1998) *Changing Private and Public Roles in Agricultural Service Provision.* ODI, Natural Resources Group, London.

Christoplos, I. and Nitsch, U. (1996) *Pluralism and the Extension Agent: Changing Concepts and Approaches in Rural Extension.* Swedish University of Agricultural Sciences.

De Silva, S. (2000) *Community-based Contracting: a Review of Stakeholder Experience.* The World Bank, Washington, DC.

Farrington, J. (1994) *Public Sector Agricultural Extension: Is their Life after Structural Adjustment?* ODI, Natural Resources Institute, London.

Feder, G., Willet, A. and Zijp, W. (1999) *Agricultural Extension: Generic Challenges and Some Ingredients for Solutions.* The World Bank, Washington, DC. Policy Research Working Paper 2129.

Rivera, W.M. (1996) Agricultural extension in transition worldwide: structural, financial and managerial strategies for improving agricultural extension. *Public Administration and Development* 16(2), 151-161.

Rivera, W.M. and Cary, J.W. (1997) Privatising agricultural extension worldwide: institutional changes in funding and delivering agricultural extension. In: Swanson, B.E. (ed.) *Agricultural Extension: Reference Mannual*, 3rd edn. FAO, Rome.

Scott, G.C. (1996) *Government Reform in New Zealand.* IMF occasional paper 140. International Monetary Fund, Washington, DC.

Walker, A.B. (1993) *Recent New Zealand Experience in Agricultural Extension.* Proceedings of the Australia-Pacific Extension Conference, Gold Coast, Australia.

White, R. and Eicher, C.K. (1999) NGOs and the African farmer: a skeptical perspective. Michigan State University, Department of Agricultural Economics, East Lansing, Michigan. Staff Paper No. 99-01.

Section One

# OFF-LOADING PUBLIC SECTOR EXTENSION DELIVERY SERVICES

Partial off-loading



Total off-loading

Chapter 2

# Chile: the Evolution of the Agricultural Advisory Service for Small Farmers: 1978-2000

Julio A. Berdegué and Cristían Marchant

## Introduction

Since 1978 Chile has experimented with different arrangements of its extension system for small farmers, all of which share the characteristic of private delivery and public funding of the services. No other developing country – and few developed countries – can show this continuous track record of more than two decades of contracting extension services with private sector organizations. Several characteristics of the extension services have evolved to adjust to the prevailing economic, institutional and political conditions in the country. This case study describes and analyses this evolution of Chile's Agricultural Advisory Services (AAS) for small farmers

AAS is used herein to describe the Chilean system. However, between 1978 and 1983, the system was known as Entrepreneur Technical Assistance Programme (ATE); between 1983 and 1997, the official name was Technology Transfer Programme (PTT), which described the overall programme which had several "modalities", each of them with a specific name such as Integral PTT, or PTT Stage I, etc. The current name of AAS was adopted in 1997.

## The Evolution of the AAS

Until the mid-1970s, Chile's extension service resembled the predominant model established in the 1950s and 1960s in most Latin American countries under the influence of USA advisors. However, in 1978 the delivery of the extension services for small farmers was privatized overnight, as part of an overall neoliberal policy which aimed at drastically limiting the participation of the government in any activity that could be conducted by the private sector. Since then, and in particular since the 1990s, many other Latin American countries have designed extension services under the influence of the so-called "Chilean model".

At least four stages can be recognized in the evolution of AAS from 1978 until 2000.

## 1978-1983: Period of Maximum Liberalization

The Ministry of Agriculture implemented through one of its agencies the Entrepreneur Technical Assistance Programme (ATE). In this programme, individual small- and medium-scale farmers could obtain fully subsidized, government-issued vouchers with which they could pay for the technical assistance provided by an independent agronomist or veterinary doctor. The individual farmer was responsible for selecting the professional that would provide the services and the farmer could terminate the contract at any time. ATE's objectives were to increase yields of basic commodities.

Underlying the system was the premise that bureaucratic interference would lead to inefficiency and that free market processes would eventually lead to an equilibrium in which the best extensionists would hold the largest number of individual contracts. In this system, the farmer was supervisor and evaluator and acted by either renewing or terminating the contract of the extensionist, depending on the results of his/her work.

This system failed due to the false assumption that there was a market for technical assistance services in the rural areas of Chile, and the system was consequently discarded. In actuality, the farmer was not free to choose between two or more professionals since in most rural areas even the presence of one qualified agronomist was rare. Thus, the farmers most often hired the one person who visited his/her farm, and the contract was established on the basis of the information that this extensionist wanted to transmit to the farmer. If the results were not those that the farmer expected, nothing happened since the cost of the service was fully subsidized and there were in most cases no other competing professionals in the area. Gómez (1991) concludes that the lack of government supervision "led to all types of irregularities and most of the times it was the peasant who was most affected...".

## 1983-1990: the Period of Maximum Uniformity

In 1983 the system was replaced by a strictly regulated Integral Technology Transfer Programme (PTTI). Supported by a World Bank loan, its design was based on the Training and Visit approach.

The PTTI sought to increase yields and production levels of the basic agricultural commodities. The PTTI was designed to stimulate the performance of "traditional agriculture" (i.e. those sectors destined for the internal market, as opposed to fruit and vegetable export production). This programme provided technological support services for small farmers and developed measures aimed at improving prices and marketing.

The system operated throughout the whole country and for all agricultural regions, although in 1987 an equivalent programme, the Basic Technology Transfer Programme or PTTB was introduced to cover very small-scale farmers, usually practicing subsistence agriculture in poor areas of the countryside.

In the PTTI, and later also in the PTTB, the government used a public bidding system to determined which private Technology Transfer Consultant firm (CTT) would provide the service in a given area. Some 92% of the contracts were awarded to small for-profit consultant firms that were comprised of one or two university level professionals who were the owners of the firm and also acted as field-level advisors to farmers. Contracts were administered by INDAP, the Agricultural Development Institute, and all aspects of the service were regulated in great detail, including the farmer/extensionist ratio (48/1 at the beginning, which later 66-72/1) as well as the types, number and frequency of the activities.

For political reasons, many qualified private organizations, such as non-governmental organizations (NGOs) and small farmers' organizations, were barred from participating in the programme. By 1989, only three consultant firms closely linked to the government held 39.8% of all contracts (Gómez, 1991). The programme reached a point at which it was highly concentrated and very rigid due to political exclusion.

PTTI worked in the sense that reasonable and timely services were received by about 25,000 small farmer households, and there is little doubt that between 1983 and 1990 the PTTI had a significant impact in terms of productivity objectives. Despite its rigidity, its "top-down" approach, and its lack of methodological concern with the issue of impact, the PTTI worked because its design, objectives and target populations were coherent with the objectives of agricultural and economic policies. Small farmers had clear and strong incentives to intensify production and raise yields, and the PTT provided the information that allowed them to respond accordingly.

**1990-1996: the "Improvement Plan"**

In 1989 after 17 years of military rule, Chile elected its first democratic government. To consolidate democracy, all branches of government were instructed to give priority treatment to the poorest sectors of the population. In less than 2 years, coverage of small farmers was doubled to a total of about 47,000 households – almost all in poor regions of the country. It was expected that PTT would extend its "productivist" objectives to these new clients; however, the gap between objectives and results soon became apparent and had a harmful effect on the system as a whole.

In response to this challenge, an "Improvement Plan" was designed in 1993-1994. PTTI's objective was changed from that of "increasing yields" to "diversifying and increasing yields". In 1994, 468 local groups out of a total of

1109 defined their main objective as the introduction of new production alternatives, and some 450 groups redirected their work to emphasize solving marketing problems. Thus production objectives were complemented by other aims, such as post harvest, value adding, marketing, access to financial services, farm management, business planning and farmers' economic organizations. Also, in 1990, the political constraints that had prevented the participation of qualified private sector organizations were removed, and the system became more diversified in terms of service providers.

## 1997-2000: the AAS

In 1997 INDAP implemented a new set of reforms that included changing the name of the programme from PTT to AAS. These reforms included: (i) a greater role of farmers' groups and organizations in choosing, contracting, evaluating and, if necessary, removing the service provider; (ii) a clear differentiation of the specific types of services and objectives to be achieved; (iii) greater responsibility of farmers in co-financing the service; and (iv) consolidation of the idea that the objective of the AAS was not simply to increase production, but to promote the competitive participation of small farmers in an open market economy. The system was to support both "hard" and "soft" (i.e. management) technologies and innovations.

In organizing the transition from PTT to AAS, INDAP recognized that many small and poor farmers practicing subsistence agriculture would have difficulties participating in market-oriented arrangements. Hence, PRODESAL was put in place. PRODESAL was implemented through a contract between INDAP and the municipal governments. The latter are the institutions that can complement agricultural services with improved access to social services of great importance to those living in conditions of poverty. By means of contracts, INDAP transfers to the municipal government about US$240 per farmer, and the municipal government uses these resources either to provide the service directly or to subcontract private consultant firms or NGOs that will take care of doing the field work. Some 20,000 of the 52,000 farmers participating in AAS fall within the PRODESAL modality.

Under this new arrangement, there is no predefined field work methodology. Local groups and their advisors are free to define their specific goals and how they plan to meet them. The field level monitoring system of INDAP focuses on results actually achieved. In this new arrangement in which farmers' organizations had a much greater say in deciding which firm should be contracted to provide the advisory services, NGOs suffered a very large reduction in their share of contracts. While private for-profit consultant firms and farmers' organizations retained their share of contracts, municipal governments became the new and very important actor in the system.

# Impact

In 1997 the Ministry of Economic Affairs jointly with the Ministry of Agriculture (Comisión Interministerial de Fomento Productivo, 1998) contracted for an external impact evaluation. It was found that on average farmers remained in the system for 5.85 years, with an average total net present cost of approximately US$4,094. The estimated total net benefits to farmers were between 0.86 and 2.14 times the net cost of the programme. The evaluation also found that there was little if any impact on those households where farm income represented less than 50% of the total household income, a situation associated with very small landholdings in regions with low agricultural potential. Thus, PTT participation had a positive and significant effect on farm revenues and total family income (Edmonds, 1998). The programme increased farm income by increasing the intensive scale of farming pursued by participants.

An earlier study by the World Bank (1994) concluded that the annual family income of those who participated in INDAP's programme in a low-productivity region characterized by a high incidence of rural poverty was US$1200 greater than that of non-participating households.

# Conclusions

Chile's agricultural advisory service for small farmers has evolved gradually over two decades. In line with the economic, political and institutional changes in Chile, the system has evolved from one promoting increased yields of basic commodities to one which is now focused on the competitive participation of small farmers' economic organizations in an open market economy.

Impact studies agree that the system has had a positive impact on key indicators of agricultural and technological performance. It has also had a positive and significant impact on the net income of the participating farmers. It appears that the system has played a very positive role in supporting the development of more competitive, diversified and productive small scale agriculture in Chile, and has had a modest but positive role in the alleviation of rural poverty.

Some of the most important design issues that have driven the changes put in place through time include:

**1.** Objectives of the system. The gradual change in objectives from a system that was concerned with increasing production to one that is now focused on the promotion of viable and competitive commercial projects implemented by farmers' economic organizations.

**2.** Role of farmers in the control of the system. A shift in the relative roles of the government vis-à-vis the farmers in determining the objectives of the programme at the field level and in selecting and controlling the private organizations that provide the service.

**3.** Content and nature of advisory services to be provided. The need to diversify the types of advisory services provided by the system as the challenges faced by small-scale agriculture became more demanding with the opening of the economy in the late 1980s. The system started providing only agricultural production advice, and it now is involved in the provision of different types of technical and professional services, including commercial, financial, farm management, post harvest, value adding and legal advice.

**4.** Role of farmers' organizations. The evolution from a system that provided services to farmers on an individual basis, to one that today works with farmers' organizations.

**5.** Heterogeneity of beneficiaries. The balance between supporting better off small landholdings with greater economic potential, and providing assistance to very poor households who practice subsistence agriculture, and the need to design differentiated services for each major group of households and farming systems.

**6.** Quality vs. quantity considerations. The trade off between number of households served and quality and intensity of the services provided to each household.

A major lesson is that an agricultural extension or advisory system takes a long time to be built. All the stakeholders (the government, the individual farmers, the farmers' organizations, the private service providers) have to learn with time, accumulate knowledge and information and correct past mistakes or to respond to new challenges. If a country is unwilling to invest in this time-demanding learning and evolutionary process, it is unlikely that it can eventually put in place an efficient and effective agricultural advisory service for small-scale farmers.

Is the Chilean advisory system sustainable? Even in a relatively long term scenario (e.g. 10-15 years) it is highly unlikely that small farmers can finance 100% of the cost without any form of government subsidy. In this sense, even after two decades and despite the significant reductions in the average cost per farmer, the system is still dependent on the annual political decision to allocate public resources to fund 85% to 90% of the programme's total cost. Since 1983 the budget of PTT-AAS has shown an average annual rate of increase of 11%, after adjustment for inflation. Over the past 6 years the annual budget of the advisory services has shown an annual increase of 4.4% in real terms, and it is likely that the current level of funding (US$22 million per year) will be maintained or perhaps even increased in the period 2000-2006.

# References


Comisión Interministerial de Fomento Productivo (1998) *Evaluación de Instrumentos de Fomento Productivo. El Programmea de Transferencia de Tecnología de INDAP.* Ministerio de Agricultura and Ministerio de Economía. Santiago.

Edmonds, C.M. (1998) Policy regimes, agrarian institutions, and the performance of smallholder agriculture in Chile: three essays analyzing longitudinal survey data on Chilean peasant farms. PhD thesis, Agricultural and Resource Economics, University of California, Berkeley, May 1998.

Gómez, S. (1991) Nuevas modalidades de apoyo a la pequeña agricultura: el caso de Chile. *Estudios Sociales* 70, 131-147.

López, R. (1996) Determinantes de la pobreza rural en Chile: programas públicos de extensión y crédito, y otros factores. *Cuadernos de Economía* 33(100), 321-343.

World Bank (1994) Chile: strategy for rural areas. Enhancing agricultural competitiveness and alleviating rural poverty. Report N 12776-CH of the Natural Resources and Rural Poverty Division. The World Bank, Washington DC, p. 54.

Chapter 3

# Estonia: the Role of Contracting for Private Agricultural Advisory Services

Ülar Loolaid

## Introduction

Estonia deserves the attention of extension professionals for its innovative solution to the development of its private advisory service and subsidy scheme. A number of different contracts have been developed for advisory service advancement. This case study describes activities and innovations introduced in Estonia since 1995.

## Challenges for Agricultural Extension Development

Estonia is a country in transition. Gaining independence, developing a market economy and building a democratic society have been the main goals of the transition process that began in the early nineties. A bottom-up principle has been the main objective for development of extension and advisory services.

Considerable investments from the governmental budget and foreign donors were allocated into the advisory service network. Three important problems led to the need for changes in the development of the extension system:

1. Delivery of advisory service did not correspond to the actual structure of agricultural production. An important segment of agricultural producers did not have access to the agricultural advisory services funded by the government. Some farmers also complained about the quality of advice they received.
2. Some of the active extension specialists were not involved in extension activities covered by governmental funding.
3. General management of extension services on the governmental level was insufficient.

## Reorganization of Agricultural Extension

The Ministry of Agriculture launched a national agricultural advisory services programme in 1995 and a new advisory subsidy scheme was introduced in 1996. Different types of contracts were designed to bring about changes in advisory service development:

1. Contracts for providing advisory services for an individual farmer approach.
2. Contracts for providing extension services for farmers in a group approach.
3. Contracts for development projects in order to provide training, support services, and extension for groups of farmers in a widespread  approach.
4. Contracts for designing the legislative framework, administration and financing in order to support delivery of advisory and extension services.

### Contracts for Subsidized Individual Agricultural Advisory Services

A model contract was designed for subsidised delivery of individual advisory service. Direct partners of the contract are legally registered farmers and certified advisers. The county administration is an indirect partner of the contract. Every county administration is responsible for the management, financing and accounting of contracts for farmers in a particular county. The contract signed by a farmer and adviser is not valid without approval by the county administration.

**Table 3.1.** Cost sharing for subsidised individual agricultural advisory service.

| Year | Contract price limit EEK* | Farmer's payment EEK* | Farmer's share % | Subsidy EEK* | Comments |
|---|---|---|---|---|---|
| 1996 | 2,200 | 220 | 10 | 1,980 | Since June 1996 |
| 1997 | 3,000 | 300 | 10 | 2,700 | |
| 1998 | 3,000 | 450 | 15 | 2,550 | Primary subsidy |
| | 3,000 | 1,200 | 40 | 1,800 | Secondary subsidy since June 1998 |
| 1999 | 3,000 | 450 | 15 | 2,550 | Primary subsidy |
| | 6,000 | 3,000 | 50 | 3,000 | Secondary subsidy |
| 2000 | 3,000 | 450 | 15 | 3,000 | Primary subsidy |
| | 6,000 | 3,000 | 50 | 2,550 | Secondary subsidy |

*1 EUR = 15,6466 EEK

Advisory subsidies enable farmers to purchase full advisory service. Farmers need to pay only for part of the service cost (see Table 3.1). According to the preliminary plan for the subsidy scheme, after 10 years the farmers should carry most of the costs of the advisory service according to the preliminary plan for the subsidy scheme.

From June 1997 until 1998, only certified advisers were entitled to deliver subsidised individual advisory service. In 1998 a standard for the advisory service was put into operation.

A model contract was designed in cooperation between the Ministry of Agriculture's Phare advisory service project and the British Know How Fund mission. County administrations, farmers' and advisers' organizations were involved in the preparation process as well. The model contract is issued, regulated and subsidised by the Ministry of Agriculture.

## Contracts for Subsidized Agricultural Extension Services through Group Approach with Decentralized Administration

A legislative framework and a subsidy scheme were designed in order to support educational events with the group approach.

Certified advisers or approved specialists are eligible to apply for funding. The county administration then makes decisions about the need for and funding of group events according to submitted applications. A wide range of participants is covered by the subsidized service since individual registration is not required.

Project organizers must provide a plan for the service activity to the county administration. After approving the plan, the county administration pays the subsidised part of the cost for the service. In line with regulation, participants must cover at least 20% of total cost of the event.

The government also finances farmers' organizations in order to provide training and advisory services for their farmers. However, this funding arrangement is not yet based on a sound contractual basis.

## Subsidized Extension Services with Group or Mass Approach or Other Support Services with Centralized Administration

Training days, seminars, demonstrations and publications are co-financed by the government using various types of contracts. One partner in this type of contract is the Ministry of Agriculture. Other partners can be enterprises, organizations, institutes, the university, etc.

The Ministry of Agriculture announces a competition of proposals every year for public financing of services. Enterprises and organizations are eligible to submit applications. The commission of extension financing is responsible for the primary evaluation of proposals and makes suggestions for financing. These types of contract are issued under general regulation of contracting for governmental funding for services.

Approved projects get twofold public funding. The first payment is made after the signing of the contract and the second payment is made after submitting the report. As a rule, the final user of provided services has to pay partially for the service.

Services necessary for general system development are contracted directly. Services financed from sources provided by the World Bank project are contracted according to regulations of the World Bank and Estonian legislation. Some examples of services for general system development are:

**1.** The Estonian Association of Rural Advisers was contracted to administer technical work for certification of advisers (governmental funding).
**2.** The Institute for Rural Development was contracted to develop a questionnaire for monitoring and evaluation of an advisory subsidy scheme (World Bank funding).
**3.** The Estonian Farmers' Federation was contracted to publish a newsletter on agriculture for farmers (World Bank funding).

### Contracts for Designing Legislative Frameworks and Financing Advisory and Extension Services

For Estonia, transition has meant legislative and administrative reforms. Designing a new law can take several years. The need for advisory system development does not fit into the frame provided by the legislation. Contracts are a suitable solution in order to create a temporary legislative framework for other contracts. The Ministry of Agriculture does not currently have legal connections with county administrations. The Minister of Agriculture has contracted with 15 county governors in order to ensure the necessary framework for a decentralised administration of advisory contracts and delivery of subsidy.

Building support systems needs more financial and technical resources than the Estonian economy can provide. The government has signed contracts with the Phare programme, the World Bank and other donors in order to ensure financing and technical assistance for building a private advisory service.

## Impact of Contracting for Agricultural Advisory Services Development

### Results of Contracting

Implementing an advisory programme and subsidy scheme in Estonia has:

**1.** Provided better access to advisory services and information for farmers.
**2.** Initiated a proactive approach of advisers in order to provide information and training for farmers.

**3.** Developed local training for advisory skills.
**4.** Certified 175 advisors.

## Impact for Individual Advisory Service

The introduction of subsidy schemes and various contracting arrangements have improved access to individual advisory service by farmers.

**Table 3.2.** Number of contracts and total cost of contracted individual advisory services.

| Year | Advisory contracts | Total cost (million EEK*) |
|---|---|---|
| 1996 | 1,789 | 3.23 |
| 1997 | 2,407 | 6.10 |
| 1998 | 2,894 | 7.95 |
| 1999 | 2,576 | 7.93 |

* 1 EUR = 15,6466 EEK

**Table 3.3.** First year percentages of contracted farmers using individual advisory services.

| Year | Percentage |
|---|---|
| 1995 or earlier | 16 |
| 1996 | 22 |
| 1997 | 28 |
| 1998 | 34 |

**Table 3.4.** Farmers' evaluation of contracted individual advisory service delivered in 1998 according to monitoring results.

| Opinion | Positive % | Negative % | Neutral % |
|---|---|---|---|
| If contracting considered farmer's interest | 93 | 2 | 5 |
| Satisfaction with individual advisory service | 91 | 3 | 6 |
| If the service was worth the contracted sum | 82 | 7 | 11 |
| Intention to continue with same adviser | 83 | 5 | 12 |
| Intention to use subsidised advisory service | 85 | 3 | 12 |
| Readiness to pay 50% of service full cost | 23 | 44 | 33 |

## Improvements Needed in Estonia's Advisory System

Access to advice on regional particularities is needed. The topics disseminated by contracted advisory services currently depend on the specialities of advisers in particular regions. The number of contracts in different rural communities varies greatly.

The advisory service is not yet driven enough by farmer demand. To a large extent, the content of the advisory service is determined by the advisers and not by the farmers. Farmers have indicated a lack of advice on economic, marketing, and legal issues, as well as issues related to joining with The European Union, machinery, construction, forestry, etc. At the same time the inservice training system for advisers is insufficiently developed, and advisers are not always satisfied with the information they can get from agricultural research.

Critical factors for the success of private advisory services development in Estonia include:

**1.** Training advisers in advisory skills and requiring training as a precondition for certification.

**2.** Initiating proactive offers of information and advice from advisers to farmers through introducing the advisory subsidy scheme and promoting incentives with project funding.

**3.** Coordinating actions for development framework and support systems for private advisory services.


## Lessons for Sustainability

The advisory subsidy was the first scheme for subsidising the agriculture sector on a contractual basis. The success of this scheme shows that the private advisory service is developed on sound principles. These principles and the positive experience of system design have provided practical lessons for development of other support systems and subsidy schemes in Estonia.

These principles may well be utilized anywhere, although policy decisions in other countries will have to take local particularities into consideration.

Different contracts reflect different objectives and, as well, respond to current needs.

Management of an advisory system has to follow changing needs and make corrections as they become necessary. Managing the system requires constant attention to identify bottlenecks and find solutions. Insufficient management capability has become a limiting factor in the development of the advisory system in Estonia.

Finally, the complexity of contracts and their interrelations is a major consideration. Ignoring these relations can lead to unsustainability. While actively engaged in contracting for extension, Estonia at present does not have an overall strategy to direct the role of contracting in the development of private agricultural advisory services; this is a major omission.

## Contacts

Hannes Aamisepp, Jäneda Training and Advisory Centre, Jäneda, 73602 Lehtse, Järva county, Estonia
Phone: 372 38 98275
Email: aamisepp@estpak.ee

Merry Aart, Estonian Association of Rural Advisers, Kreutzwaldi 1-71, 51014 Tartu, Estonia
Phone: 372 7 421718
Email: merry@online.ee

Kristel Jalak, Agricultural Information and Registry Centre, Kreutzwaldi 1, 51014 Tartu, Estonia
Phone: 372 7 421595
Email: kristel@reg.agri.ee

Olav Kreen, The Ministry of Agriculture of Estonia, Lai 39/41, 15056 Tallinn, Estonia
Phone: 372 6 256172
Email: olav@agri.ee

Kaul Nurm, Estonian Farmers' Federation, Teaduse 1, 75501 Saku, Harju county, Estonia
Phone: 372 2 721 783
Email: etkl@online.ee

Ruve Šank, Certified in 2000 during preparation of this case study.

Hanna Tamsalu, The Ministry of Agriculture of Estonia, Lai 39/41, 15056 Tallinn, Estonia
Phone: 372 6 256236
Email: hanna@agri.ee

# Chapter 4

# Federal Republic of Germany: Contracting for Agricultural Extension in Thuringia

Jochen Currle, Volker Hoffmann and Andrew D. Kidd

In Germany, each of the 16 Federal States is responsible for agricultural extension. This means that the organization of extension is quite diverse. Major differences exist in terms of both the financing and the delivery of extension among the states (Hoffmann *et al.*, 2000). This case study deals with the State of Thuringia, one of the five new German states that joined the Federal Republic of Germany in 1989 after the fall of the Berlin Wall. In the past 10 years there have been a number of changes in extension. For instance, a significant proportion of the state budget for extension has been contracted to private extension companies since 1998. However, the state has remained involved in delivering more public good aspects of extension and support services, such as applied research, agricultural education, and counselling on farm crisis and environmental issues.

## The Farming Sector Context

At present, some 4,300 individual farmers and farming enterprises cultivate 802,000 ha of arable land in Thuringia state. The large production units of the socialist era still dominate the farming sector, with some 608,000 ha (76%) being cultivated by 415 farming enterprises with more than 500 ha each and 160,000 ha (20%) by 916 farm units between 500 and 50 ha each. The remaining 34,000 ha (4%) belong to 3,013 farms of less than 50 ha each.

The largest land holdings are owned by either cooperatives (207 with an average holding of 1,571 ha) or private companies (GmbH; 226 with an average holding of 755 ha). Most farms of less than 500 ha are either owned as partnerships (GbR; 258 with an average holding of 318 ha) or as family farms (806 with an average holding of 117 ha). There are also a large number of part-time farmers holding some 3.3% of arable land with an average size of 11 ha.

Practically all farms over 500 ha are privatized, successors to the former socialist state production units. A large number of smaller farms were re-established by private owners whose previously confiscated land was returned to them.

## Recent History of Extension in Thuringia

In the early 1990s the dramatic structural changes in the farming sector brought much uncertainty and serious need for advice on legal and structural issues. But during the first 2 years after the unification of Germany there was no official extension structure in Thuringia to provide sound advice. Several incompetent and fraudulent private extension companies took advantage of this situation, which roused the state agricultural administration to establish an official extension body in 1991. Thuringia adopted a system similar to that used by the state of Bavaria to its south (Hoffmann *et al.*, 2000). The system was publicly financed and delivered through 12 state agricultural offices employing about 80 agricultural advisors.

Seven years after unification, in 1996, the level of uncertainty in the sector decreased and the need dwindled for legal advice and assistance with company plans for the purpose of gaining credit. Requests for specific economic and production advice increased, presenting a challenge to public advisors to reorganize the advisory service.

In the midst of slowly adapting to the changing needs of their clients, the agricultural administration was ordered by the state treasury in late 1997 to cut 120 staff positions. The state agricultural extension system was an easy target. However, the idea of establishing a private extension organization using public advisors failed since the majority of these advisors decided to remain civil servants, though this often meant changing positions and being assigned quite different tasks and functions.

## Contracting Placed on the Agenda

During this time the state decided to use a significant proportion of its extension budget to contract for services. Contracts were developed, independent of any particular extension task and aimed in part at fostering the establishment of private organizations for extension in Thuringia. DM1.4 million per year were provided to cover advisor personnel costs. About 60% of this total originated as a grant from the European Commission with the remainder coming from state finances.

Private companies generally applied for broadly framed contracts to "give support for individual farms in the areas of technical, economic and financial, and administrative farm management". Their objectives fell basically into the realm of "private good" extension services.

Stipulations in the contracts between the state's Ministry of Agriculture, Nature Protection and Environment, and each private extension company were several. The personnel cost subsidy was to be provided for certified advisors only, i.e. those who carry out full-time advisory work, are based in Thuringia and provide advisory work within state borders. The contract arrangement was fixed at a set rate of up to DM103,000 for each certified advisor over a period of 3 years. Over the 3-year period, the amount released for the first year could reach up to DM50,000, DM30,000 in year two, and DM23,000 in year three. The terms of the contract were monitored mainly through an annual reporting schedule that each advisor was required to submit to the state.

The state adopted its new functions of regulation and quality control of the private sector by requiring the certification of agricultural advisors, based on their professional qualifications and experience. Passing a capability assessment administered by representatives of different agricultural institutions (ministry, public research institutions, farmer association, horticulturalist association, and ecological farmers' association) was a precondition for being placed on the certified list. Advisors were also invited by the state agricultural administration to attend cost-free training given by public agricultural research institutions as well as courses on new rules and regulations for the agricultural sector.

The emergence of contracting for extension services also resulted in some costs being transferred to farmers or farm enterprises in the form of advisory fees. The rate was not set but averaged about DM66 per hour. At present, about 50 extension companies with a total of 68 certified advisors provide agricultural extension in Thuringia. One of these companies is owned by the state branch of the farmer association, to which every farmer must belong. This company, however, operates largely independently of the association and on similar terms as other companies.

The state ministry continued to maintain a much reduced extension service: 15 positions out of the former 80. In line with the legal requirements stated in the new extension policy of 1998, this service focused on "public good" services. The advice covered mainly farm crisis counseling, environmental issues, promotion of women and families in rural areas, plant protection, and human nutrition in rural areas.


## Impacts of Contracting Out

The Thuringia experience of contracting out to support the privatization of extension is still relatively recent, although some clear impacts can already be seen.

## Rapid Adjustment to Changing Needs

The contents of advisory work have been adjusted rapidly in response to the changing needs of farmers and farm enterprises, and farmers only pay for what they need. It is highly unlikely that the previous public extension service, that used to give advice on legal matters and development plans and was couched in the security of the civil service system, would have been able to respond in the same way.

## Reduction in the Demand for Extension Among Smaller Farm Enterprises

The transfer of part of the cost burden to farmers led to a dramatic decrease in requests for extension. In the last year of the free public extension service, some 80% of the farmers sought advice in one form or another. With the introduction of private extension in January 1998 this figure fell to about 13% (731 farms). Yet half of this number was made up of the larger farms of more than 500 ha. Hence, about 88% of larger farming enterprises paid for advice, whereas the figure was only 9.3% for farms smaller than 500 ha.

The fees (DM66 per hour) were perceived as being relatively steep, considering the high subsidy provided through a contractual arrangement with the state. It appears that the private extension companies did not pass on this subsidy to their clients, but used it for capital investment in offices and infrastructure.

Indeed, the fall in demand in Thuringia led to an effective over-supply of advisors. As a consequence a good number of certified advisors began to work outside the state and some outside the field of extension. This contravened the contract stipulation and gave an unexpected headache to the state agricultural administration. Regulation and administration costs rose as a consequence.

## Reduction of Public Costs

Since the main driving force for contracting out was fiscal prudence, did the state save money? In budgetary terms the figure was reduced from DM5.2 million per year to DM2.5 million per year (including residual public service, contracting out costs and administration). This represented a decrease of over 50%.

## Sustainability and Replicability

*Sustainable Advisory Offer for Larger Farms*

Two positive outcomes favour sustainability and replicability of the approach taken in Thuringia. First, public expenses for the support of agricultural extension have been cut considerably. Second, the advisory services available to the farmers who are willing and able to pay is more than sufficient, with a level of competition that supports quality control through market mechanisms.

None the less, the usefulness of contracting arrangements that are independent of specific extension tasks is clearly questionable. An imbalance in the use of extension services is clearly apparent to farmers. The Thuringia model tends to exclude the majority of small- and middle-sized farms from extension services. The maintenance of public extension services for farms in crisis does nothing to counterbalance this situation, for experience shows that in most cases the service is called on too late and the farms go under.

To give the agricultural administration in Thuringia its due, it realized that many farmers were not being served through its contracting arrangements and that costs for regulation and administration were too high. Thus, the state changed its policy at the beginning of 2000 and has adopted a system used by the neighbouring state to the north, Saxony-Anhalt. The state now provides support to agricultural extension in a task dependent manner. The cost of extension will be reimbursed to farmers up to 60%, but not more than DM1,200 for a single farm. The contracting programme between the state and private extension companies is set to phase out within the coming 3 years.

*Conditions for the Replication of the Approach*

Budget cuts – the main driving force for change – did not come from within the state agricultural administration, but were imposed by a cost-cutting treasury. It is unlikely that contracting would have been put on the agenda without such external pressure.

The Thuringia model of contracting out is best seen as a transitional programme, where the private sector is poorly developed. State support gives private extension companies a 3-year period to develop perspectives, to build up working contacts, get acquainted with the specific situation of farms and establish office infrastructure. Without the state reducing the risk of market entry it is unlikely that such a number of private extension companies would have been established.

In addition, financially strong farm enterprises willing and able to pay for services were necessary to maintain even the reduced level of demand. If the land holdings had been more evenly distributed then the overall demand may have been still less and the private sector could not have developed as it did.

An important precondition for the development of the private sector was the availability of well-educated agricultural experts who could take over the task. As the Thuringia case indicates this is not effortless in a transitional economy where free public service extension previously had a monopoly. Contrary to what was planned, public service advisors largely shunned the potential insecurity of the private sector. Thuringia had to rely on a new pool of both young local professionals and advisors from other regions where private extension was already established. This required giving special attention to their professional development.

# Lessons Learned

## Task Independent Financing – a Second Best Solution?

Two reasons led to the decision of agricultural administration to support private companies financially over three years of establishment. The first one was a sense of urgency following the order in late 1997 to dismantle much of the official extension service in January 1998. The other reason for taking that measure was a European Commission rule (VO (EG) No. 2328/96) that restricted the use of direct reimbursement of extension costs as a financing mechanism.

Although the contracting system was not highly regarded, it has been able to assist in the development of a good number of quality private service providers. But regulatory and quality control activities associated with contracting created considerable transaction costs. Furthermore, the contracting arrangements were rather vaguely stated and were not directly linked to any specific extension tasks or targets. The arrangements lacked a certain amount of transparency and allowed for an oversupply in extension provision. Without taking specific measures to draw more farmers into the new arrangements, a reduction in the number of private extension companies was inevitable.

## The Role of the State – Still an Important Player

Even though the core aspects of agricultural extension have been passed over to private providers it is clear that the state still has an important role to play in guaranteeing that quality advice remains available for all farmers in all relevant areas. In Thuringia, the state considers quality management as one of its core functions. Certification is an important aspect of quality management. Monitoring through the use of annual reports is likely to become less useful as quality control mechanisms become more market-based rather than hierarchy-based. This is also suggested by the longer experience of Saxony-Anhalt with reimbursement schemes (Grygo, 1996).

The Thuringia case suggests that there are aspects of extension that are not readily absorbed by the private sector as they are more "public good" in character. These include advice on environmental issues, such as the use of plant protection pesticides or ground water protection, and farm crisis counselling. From 2000 onwards farmers can opt to call on a private advisor and get reimbursed up to 90% of the cost. Nevertheless, there will remain areas where public interest may conflict with private interest. For instance, requests for advice will hardly arise from the farmers' side when even a nominal payment is involved.

In contrast there are situations where the need for advice is duly felt and articulated by the farmers. In such cases, there is little need for the state to provide advice but only to enable farmers to obtain that advice. This may involve some basic quality control and targeted financial support – depending on the issue, the situation and the need of the farmers. As in all cases of public sector, clear political will and direction are necessary preconditions.

**Collective Action Sector – Only Marginally Developed**

It is striking that in comparison to the "old" states of Germany (i.e. former West Germany) there is little farmer-organized extension in Thuringia. Though present regulations do not exclude farmer organizations from obtaining direct support for extension, strong farmer-determined constructions have not emerged. This may have something to do with the history of agriculture under socialism and the more recent dramatic transformations in society that have supported building the necessary communication ties among recently established farms. Institutions of civil society are generally much weaker in the new states of the former East Germany.

Elsewhere in Germany change is also taking place in the institutional arrangements of advisory services (Hoffmann *et al.*, 2000). The establishment and development of farmer extension groups in several German states has been promoted, heavily supported by the state agricultural administrations in their moves to reduce reliance on publicly delivered extension. Farmers, accustomed to a public extension service may need time and support to organize themselves and create farmer-determined institutional arrangements for advisory services. This is an issue that still needs to be on the agenda in Thuringia.

# References

Grygo, H. (1996) Landwirtschaftliche Beratung – quo vadis? *Ausbildung und Beratung* 5/96, 88-89.

Hoffmann, V., Lamers, J.P.A. and Kidd, A.D. (2000) *Reforming the Organization of Agricultural Extension in Germany: What Lessons for Other Countries?* AgREN Network Paper No. 98. ODI Agricultural Research and Extension Network, London.

Lanz, M. (1993) Gruppenberatungsansätze im Agrarsektor. *Berichte über Landwirtschaft* 71, 98-105.

Chapter 5

# The Netherlands: Going Dutch in Extension, 10 Years of Experiences with Privatized Extension

Jet Proost and Paul Duijsings

Ten years ago the Dutch government privatized its national agricultural extension and research services. The extension service was renamed DLV and in 1990 became independent from governmental funds. With the introduction of commercialization and competitiveness in the Dutch agricultural knowledge system, both advisors as well as end users of information and knowledge were affected.

## History

More than a century ago prices of agricultural produce dropped dramatically due to cheap imports from the United States of America, which created a deep crisis in Europe. Several countries sought to protect their home markets, but the Dutch government decided not to, deciding instead to reinforce the agricultural sector by investments in an agricultural knowledge and information system of extension, research and education, better leasehold legislation and land improvement. There was no protectionism nor income support for Dutch farmers, but rather a century of large investments in a knowledge infrastructure. This approach explains to a large extent the success of Dutch agriculture.

During this period, a very high use of inputs created environmental problems like nitrate concentration in ground water, toxic wastes from agro-chemicals and a surplus of manure. Increasing public pressures resulted in stringent policies and regulation restricting agricultural production. A public extension service could not survive in such circumstances and in the late eighties the government had to reconsider its position (Tacken, 1991). The argument was that since farmers were making a living like anyone else, why should they be supported with free extension and research? In the early 1990s food supplies were abundant, there was no strategic argument to support farmers and to safeguard consumers against food shortages, nothing like the country experienced in the forties, during and after World War II.

From 1985 on, an internal working group of the Ministry of Agriculture worked on a reorientation and a new mission-statement for the extension service. The group developed a number of alternative structural and financial models (Tacken, 1991). Discussions on reducing the state budget involved considerations on ideology as well as the empowerment of farmers and other practicalities. Farmers' organizations liked the idea of more influence, but were troubled by the idea of co-financing. Finally in 1988 government and farmers' organizations decided to disconnect extension services and policy affairs. Three main tasks were to be carried out: (i) 12 provincial teams were to explain and promote agricultural policy; (ii) two information and knowledge centers were to prepare policy decisions and provide a liaison between research institutes and the extension service; and (iii) a national extension service was to be fully privatized in 10 years.

Five major changes (Bos, 1989) resulted from the reorganization:

**1.** Better delineation of tasks in and between the various parts of the Ministry of Agriculture.
**2.** Clarification of the position on the new parts of organizations in the knowledge network.
**3.** Adjustment or modernization of the organizational structures and working methods.
**4.** Where necessary, combining forces in the larger parts of the Ministry, in line with development of more intensive cooperation between different disciplines.
**5.** The start of a change in a corporate culture in various parts of the Ministry as a result of the change in task division, other attitudes and working methods.

The intentions were good, but the new structures did not last very long. The provincial teams were abolished in 1993, and in 1994 the Ministry of Agriculture opened five regional affiliations. The Information and Knowledge Centers endured a long series of reorganizations and in 1998 were finally amalgamated – a small policy advisory body remains.

For the DLV the first years of the privatizing process with decreasing governmental funding (10% per year) were rather difficult, in the sense that the organization did not have full freedom in its operation. Its Governing Board with 50% of its members from the Ministry of Agriculture DLV was still bound to ideas and developments orchestrated by the government. This situation, at least in the short run, meant job security for the employees. However, as DLV decreased its overheads by 40 to 50%, and proved to be more cost-effective, it wanted to be responsible for its own financial affairs.

In 1995 an agreement was reached which divided public funds for extension in a different way. From 1996 until 2000 only DLV and LTO, the national farmers' association, were allowed to apply for the 30 million Dutch guilders subsidy per year from the government for the implementation of agricultural projects. LTO started its own extension service on a commercial basis in 1996. The government no longer supported organizations financially, but provided funds on a project basis. Both LTO and DLV submitted project proposals once a year to address key issues according to a plan set up by the Ministry.

DLV started off with a sectoral organization with 12 business units. Soon more activities were added to the business and new groups of clients were served, i.e. other actors in the production and transformation chains besides farmers. Because of developments in agriculture, a broadening of activities took place in rural development and food chain processes. While the relation with the government had to be loosened, relations with others had to be built up. DLV also developed activities outside The Netherlands in Belgium, Germany and Estonia, and has an increasing network of foreign offices and subsidiaries. The DLV Advisory Group (Advies Groep) is a full-service advisory agency employing 900 people in 25 branches in The Netherlands and abroad.


## Present Situation

Because of the change from collective financing into a 100% market operation, the management, structure and culture of DLV have drastically changed. Some of the most outstanding changes are the following:

### Client Orientation

At present the focus is on the client and not on the advice. In the 'old' situation (before 1990), the TOT (Transfer of Technology) model was quite popular, with agricultural researchers developing innovations which were then transferred to the farmer by an extension agent. In the present situation an advisor brings only information that is asked for by farmers and works according to a contract. So the client's question is central in the advisory process.

### New Skills

The implication for advisors is that they must master new skills, like selling, but even more importantly the ability to clarify a client's problem or question. Much depends on the capabilities of the advisor, to understand a farmer, to oversee the situation, to comprehend the personality of the client. This includes good listening skills, asking the right questions, scrutinizing facts and figures, and counselling.

### Planning

Another change involves writing reports that often include the advice and further steps to be taken. In the case of complex issues sometimes a colleague is asked for an internal quality check.

**Monitoring Clients**

If the client is the centre of activities, monitoring of clients' appreciation is of vital importance. An advisor has to verify whether the service met expectations, but the organization also has to check if the service was up to standard. Under the slogan "we are satisfied if you are satisfied" questions are asked such as: Did the advice answer your question? Did you follow the advice? How satisfied are you with the advice?

The administration has become more complicated over the years, because all sorts of clients' data have had to be recorded, including information about advisors' activities. The professional capability of an advisor is measured against two indicators: the bill turned over and the clients' satisfaction. Until the beginning of the year 2000 each advisor had his or her individual target. This system was replaced by group targets, which enables the whole team to divide the work on the basis of a yearly plan and then calculate their annual income. The buzz word of the new millennium is social cohesion, meaning: we need each other's capabilities. In that sense individual competition can frustrate the process of development and quality improvement. Over the past 5 years emphasis has shifted from rate of return to quality of the advisory process.

More and more, DLV has to operate in a market where the number of farmers is decreasing and competition is getting stronger. Many organizations like suppliers, private consultants, and cooperatives have their own systems for advising clients. The costs are often calculated in the price of the produce. The LTO Advisory service, the former extension service owned by the Farmers' Organization, is a strong competitor. In such a competitive situation quality is an important way of distinguishing the companies' services.

Privatization set off a chain reaction within the agricultural knowledge and information system. The word "extension" tends to be exclusively used for transfer of information from government and "advising" is used for giving service to clients who take the initiative to request it. The "supply" of information under the governmental extension service has gradually changed to a situation of "information on demand". Farmers can look around for the best price and buy the information wherever they want. But farmers need to have a clear question in mind and know what kind of information they are looking for. They also need to have an overview of the available sources of information. And lastly they must be able to pay for it.

# Contracting

An individual contract is made for every request for advice, containing the name of the client, description of the assignment, relevant delivery time, payment agreement and employee(s) who will fulfill the assignment. Both the client's signature and the advisor's are required. Depending on the question and services needed this can be a simple order confirmation or a more elaborate agreement.

The reasoning behind this procedure is to prevent differences of opinion between client and advisor that can lead to insufficient quality and payment problems.

In the case of a larger project a more elaborated agreement is made, including planning of the project, results, timetable and financial arrangement.

In the case of an individual farm business the contract is drawn up by the advisor together with the client, and a yearly plan is set up with concrete, measurable and realistic targets. This plan indicates the direction of the advising activities throughout the year, and the progress in achievement of results which is to be continuously monitored.

This particular approach to contracts may seem rigid, but the reason for it comes from DLV's quality policy. In surveys, clients indicated that DLV insufficiently controlled the process and didn't have a proper overview of the results expected. After the introduction of the contract system, fewer complaints were made and there was less refunding.

Another very practical reason was the state of financial affairs of the company. In a situation without subsidy from the Ministry as a stable source of income for exploitation, it is very important to have regular revenues and to keep the time between delivery of the advisory service to the client and payment to DLV as short as possible. A deficit in the balance of payments could jeopardize the organization's liquidity position. The contracting system enables sending divided invoices during the advisory process.

Setting up a contract helps the advisors in their professional development. By making a contract they have to analyse and describe accurately the client's need for advice. Furthermore the advisor has to assess his own capabilities in relation to the client's needs by using this system also the planning of activities and respecting deadlines improved considerably.

There are three critical factors for the success of contracting: (i) aiming for achievement of results, performance monitoring and control by management; (ii) tight management on accomplishment of agreements with the clients; and (iii) the client is always right, and without a signed contract an advisor loses his right of appeal.

## Support for the Advisory Process

In order to make the advisory processes successful, several supportive processes are needed, viz. development of information products, exchange of knowledge and information, and recruiting and selecting new employees (of the initial 900 employees only 245 remained).

A client-oriented attitude requires continuous renewal of products. Looking into the future and forecasting tomorrow's needs is one of the hardest aspects of this support function. The supply of knowledge and information is now partially coming from public research institutes. Other sources of information are farmers themselves. Brainstorm sessions are organized with the top ten business farmers. Also, farmers' complaints are reviewed and studied. Market panels are a useful instrument and serve as an exchange with clients about their experiences with DLV. Within DLV, employees with ideas are provided facilities to develop them and can organize 'think tanks' to work out ideas. DLV conducts its own search for new products worldwide, through its many international contacts. Also signals from trade and manufacturing industry are continually scrutinized.

The survival of a knowledge intensive organization like DLV largely depends on its ability to make the right information available to the right employees at the right time. A large proportion of the advisors work from home. Contact through the DLV Intranet enables them to obtain the latest information and to communicate quickly with headquarters and with their colleagues. Knowledge management has become one of the leading functions within DLV.

To develop quality improvement at each level of the organization, selective changes in personnel were instituted. In recruiting new personnel, commercial and social skills are important. The human resource manager must differentiate tasks by competence. Building relationships with clients has become the most important asset for all employees, rather than salesmanship.

## Effects of Privatization

Several effects of the privatization can be distinguished, according to Duijsings (1998), Tacken (1997), Proost and Matteson (1997), and Proost and Röling (1992). DLVs independence reinforces the effectiveness of the advisory work; advisors are truly independent from the government, and their work is no longer identified with the interests of the state.

Farmers can choose the best available source of advice. They are more particular about quality, and they expect information from advisory services to be clear, practical and applicable immediately. Farmers are looking for custom-made knowledge and information that fits their farming situation and specific needs.

Money increases the effectiveness of the information, although no large-scale research has been undertaken so far. But farmers report that they treat the information in a different way, now that they have to specify their question to the advisor and have to pay for it. Privatization also has effects on other suppliers of knowledge and information in the agricultural sector; financial arguments are dominant over administrative. Of course, competition may lead to a spectrum of different information sources, and also to doubling of efforts. One criticism of the current system is that the agricultural knowledge and information system in The Netherlands, which has always been characterized by an openness of information flows and strong linkages between actors, has become less so as a result of the process of privatization and commercialization. Another criticism is that farmers in remote areas, those growing a minor crop, or those who cannot afford to pay for information are not served by the privatized organizations.

Privatization has, however, provoked all sorts of initiatives by farmers' organizations. Some have created an information service for their particular conditions. Others, like growers' associations, employ their own advisor or develop Internet sites.

In the transfer of knowledge and information, topics for which no interest is shown from the side of the clients will disappear from the 'menu', although they may be relevant for society as a whole – like some of the environmental issues in The Netherlands. In short, privatization has had a major influence on agricultural targets, target groups, issues, methodologies and exchange of information. Money has become the decisive factor.

# References


Bos, J.T.M. (1989) Privatization and reorganization of the Dutch agricultural extension service. In: *Agrarische Voorlichting*, No. 5, May 1989. Ministry of Agriculture and Fisheries, The Hague, The Netherlands.

Bos, J.T.M., Proost, M.D.C. and Kuiper, D. (1991) Reorganizing the Dutch extension service: the IKC in focus. In: Kuiper, D. and Röling, N.G. (eds) *The Edited Proceedings of the European Seminar on Knowledge Management and Information Technology.* Wageningen University, Communication and Innovation Studies, pp. 65-73.

Duijsings, P. (1998) Information and communication management in a privatised extension organisation: the case of the DLV. In: *Assessing the Impact of Information and Communication Management on Institutional Performance.* Proceedings of a CTA workshop, Wageningen, The Netherlands, 27–29 January 1998, pp. 59-68.

Proost, J. and Röling, N. (1992) Is privatization the answer? 'Going Dutch' in extension. In: *Interpaks Interchange.* University of Illinois at Urbana Champaign. Vol. 9, No. 1, pp. 3-4.

Proost, J. and Matteson, P. (1997) Integrated farming in the Netherlands: flirtation or solid change? *Outlook on Agriculture* 26(2), 87-94.

Tacken, W.A.G. (1991) Why and how did we change our Agricultural Advisory Service Approach in the Netherlands? In: *Challenges and Threats for Agricultural Advisory Services Created by Global Market Situation and New Emphasis on Rural Areas.* Proceedings of a seminar in Helsinki, December 1991. University of Helsinki and the Scandinavian Association of Agricultural Scientists, pp. 46-57.

Tacken, W.A.G. (1997) Landbouwvoorlichting, wie betaalt? In: *Proceedings of the International Agriculture Day 'De prijs van landbouwvoorlichting'.* Amsterdam November 1997, Royal Tropical Institute Amsterdam and Royal Agricultural Association Wageningen, pp 23-32.

# Section Two

# CONTRACTING TO PROMOTE ENVIRONMENTAL SERVICES

# Chapter 6

# Australia: Contracting to Prevent Land Degradation: Landcare, a Success Story

Trevor Webb, John Cary and Andrew Campbell

## Identification of the Case

Landcare is a uniquely Australian form of participatory natural resource management primarily focused on productive agricultural landscapes. The community landcare group forms the fundamental unit that underpins the landcare movement. Such units comprise a group of people, usually farmers, who are concerned about land degradation on their properties and their immediate surroundings, and who are interested in working together to ameliorate or repair the degradation. Landcare groups have been particularly successful in raising awareness of land degradation in rural Australia and their acceptance among rural Australians is evidenced by the growth of groups from less than 100 in 1985 to nearly 5,000 in 1999.

This case study focuses upon one such group, the Tod River Catchment Landcare Group, based around the Tod River Catchment on the Lower Eyre Peninsula in South Australia. The funding of projects implemented by landcare groups comes largely from the Federal Government's Natural Heritage Trust administered through partnership agreements with the States and Territories of Australia. But first, it is essential to outline the broader national and state context in which the landcare group operates.

The Federal financial support of community landcare groups commenced in 1984, though substantial resources were not accorded to landcare groups until the 1990s, declared as the Decade of Landcare, and the establishment of the National Landcare Programme. This programme established mechanisms that facilitated the disbursement of grant monies to community groups to address land degradation. During the mid-1990s the Federal Government established the Natural Heritage Trust (NHT). Partnership Agreements established between the Commonwealth government and the six States and two Territories set out the terms and conditions upon which Commonwealth funds are provided. They define the natural resource management outcomes sought, activities to be undertaken, any institutional reforms required and performance indicators. Project funding arrangements encourage the integration of individual projects into catchment or regional natural resource management (NRM) plans. Partnership agreements require that in considering funding of activities on private land the amount of public benefit received relative to private benefit must be taken into account. Any direct beneficiaries of funding are expected to make some contribution to project funding.

In 1997 the South Australian (SA) Government through the Department of Primary Industries and Resources (PIRSA) and the Department for Environment, Heritage and Aboriginal Affairs committed resources for the development of a regional approach to natural resource management. In 1997/98 the SA government received NHT funding to match their contribution for a project entitled *Accelerating Regional Implementation of NHT in SA* (ARI). The ARI project was a devolved grants scheme that aimed to provide coordination and technical support for a number of projects that would accelerate the implementation of on-ground works to tackle land degradation. Studies were undertaken adopting a regional approach which described problems, proposed works and identified potential pilot projects. Criteria against which potential projects were identified included the nature of the land degradation issue, the public benefit in addressing it, and the existence of a committed community group capable of assuming responsibility for the management and implementation of the project.

Where such conditions were met, funds were disbursed to the relevant community group through a financial contract between PIRSA and the group, which then managed the project locally. At the local level, on-ground works were implemented through the payment of incentives to individual landholders and managers. The incentives were based on an agreed cost-benefit framework for the overall ARI project, though there was some latitude for local modification to achieve particular aims and objectives. Expressions of interest from farmers to carry out works were called for and, subject to assessment against regional priorities, funds were allocated to works on individual properties. A part payment was made to the farmer and the remainder of the incentive was paid following the audit of the completed works. An agreement between the farmer and the incorporated community body provided the basis for the transfer of monies between the two parties.

The Tod River Catchment Salinity and Water Quality Management Project was one of the projects funded under the ARI initiative, and the Tod River Catchment Landcare group was responsible for the administration of the project. Agricultural landuse in the Tod River Catchment is primarily focused on broadacre cereal cropping and sheep grazing. Approximately 54 landholders have property within the catchment. Some of the catchment is subject to waterlogging, while a significant proportion is subject to salinisation, soil and gully erosion and soil acidity. These problems not only detract from the agricultural potential of the area, but also reduce water quality. This is important as the Tod River Catchment is the only surface water resource on Eyre Peninsula and is used as a community water supply following its collection in the Tod Reservoir.

The Tod River Catchment Landcare Group formed in 1989 and has taken part in a range of activities primarily associated with the management of salinity in the catchment. This included trial revegetation sites, drainage of saline sites, stubble retention/reduced tillage field days and liming and use of deep rooted crops to increase the efficiency of water use on cropping lands. The landcare group also carried out a survey of farmers in the catchment to identify what they saw as their major land degradation issues and concerns.

The suggestion that the Tod River Catchment Landcare Group take on a pilot project as part of the ARI initiative was supported at a meeting of local landholders. The broad objective of the project is to decrease water and soil salinity in the Tod River Catchment, and thus improve water quality in the Tod Reservoir. Additionally the project sought to encourage various farming practices within the catchment to assist in salinity control. Works proposed in the project include fencing and revegetation of creeklines, establishment of improved pasture for both saline and non-saline areas, woodlot establishment, application of lime and gypsum, fencing to protect remnant vegetation and revegetation of recharge areas. The cost-sharing arrangement saw the provision of funds, up to a maximum of a 50% contribution, based on whether the works have largely on-site benefits to the landholder or largely off-site benefits to the general community.

Letters were sent to all landholders in the catchment explaining the project and seeking interest in carrying out on-ground works. Landholders were informed of the cost sharing arrangements for each type of work. A Steering Committee of eight individuals was established and constituted as a sub-committee of the landcare group. The Steering Committee employed a project officer for one day a week to coordinate the project. While the project was implemented as a community project, the State Government provides considerable technical support through PIRSA. Expressions of interest from landholders for PIRSA officers through field sites and modifications and improvements assessed on-ground works suggested.

Approximately 50% of the farmers of the catchment expressed interest in carrying out on-ground works. Landholders agree to carry out particular works and sign a "Works Proposal" which details the works intended, the location of where the work will take place, and the contribution from the landholder and that from the government through the landcare group. The on-ground works must be completed within 1 year of approval and funds are paid on inspection at the completion of the project. Landholders are required to maintain records of expenditure in relation to the project for its administration. While the landholder is under contract to the landcare group for carrying out the works, the landcare group itself is bound by an agreement with PIRSA, which in turn is bound by the Natural Heritage Trust Partnership Agreement with the Commonwealth Government.

In such projects across Australia, the signed agreement between the landcare group and the individual farmer is usually very short and practical, of one or two pages, stating what works are to be undertaken and how much the respective parties are to contribute. It is more a documentation of the good faith between the group and the farmer (a "written handshake") than a watertight legal contract. The legal accountability in effect rests with the landcare group through its contract with the state government agency. Landcare groups manage their risk through the practice of withholding most of the payment until the work is completed to their satisfaction. Nevertheless, the requirement for landcare groups to become incorporated legal entities for such purposes, and the attendant administrative demands of managing sometimes large amounts of public money (up to $0.5m per annum), is an issue which has concerned voluntary landcare groups since the inception of the National Landcare Programme. It is a major reason why so many groups apply for funding to employ a coordinator, either full-time or part-time.

## Impact

The project was well received by local landholders, of whom approximately half committed to perform some on-ground work. The most common incentive taken up by farmers in the catchment involved liming and gypsum of soils. These offered a direct benefit to the landholder by increasing yield in addition to reducing water recharge and thus contributing to salinity reduction. In recognition of the lower level of public benefit compared to private benefit, farmers carrying out these activities received less financial assistance (25%) than other activities with far greater public benefit such as revegetation of recharge areas and creeklines, which received a 50% NHT contribution. There was also increased fencing along creeklines to exclude stock from riparian zones which assists in improving water quality and also contributes to a reduction in soil erosion.

A number of factors have contributed to the success of the project. The work the landcare group had already carried out in the catchment, in particular the survey of landholders' concerns, provided an excellent source of preliminary information from which the project could be developed. The established reputation of the landcare group gave the project credibility and local ownership. The part-time project officer relieved community volunteers of administrative paperwork, and facilitated access to the technical support of state agency staff, which was significant. Sufficient funds were available to meet landholders' requests.

The focus on the Tod River Catchment was logical in a natural resource management sense, though biophysical definition of a catchment can be difficult to translate into a bureaucratic or legal framework. Farm boundaries rarely coincide with biophysical definitions. Thus individual landholders may hold contiguous lands that extend over the boundaries of the catchment, or hold noncontiguous parcels of land both within and outside the catchment. One of the constraints of receiving funding from the ARI project was that the funds were to be spent only inside the catchment. This created some difficulties where farmers who were enthusiastic about the project were only able to obtain funding for on-ground works within the catchment, despite their enthusiasm for taking out similar on-ground works on other parcels of land they managed but that were outside the catchment.

Changes in land management practices upstream may have an impact upon landholders downstream. For example, the reduction in aquifer recharge upstream, while improving water quality in Tod River through a reduction in salinisation may lead to an overall reduction in stream flow. Thus, some landholders downstream expressed concern over the impact of reduced water availability on their productivity. Providing information about the consequences of implementing particular practices was fundamental to the broad acceptance of the project.

## Sustainability and Replicability

Some of the changes brought about through the project are likely to be sustainable, though the impact of others will be shorter lived. Thus activities such as revegetation and fencing will provide on-going and long-term benefits to the sustainability of the catchment. In contrast, activities such as liming and the addition of gypsum will have an immediate impact, but the benefits will decrease with continued cropping and grazing. Similar actions in the future may be required to maintain these benefits. In such cases there may be hesitance on the behalf of farmers to perform independently such on-ground activities, in which case the catalytic impact of this project will have been negligible. However, the private benefits gained through increased yield may be enough to encourage farmers to continue pasture-improving practices.

The contractual arrangements in this case depend on the existence of credible social and institutional structures. Clearly, for any devolved grant scheme to work there needs to be some organization which is able to take up the responsibility for managing the project. In Australia the extensive network of community landcare groups has been able to take this responsibility. Importantly, community landcare groups build upon local support and enthusiasm for dealing with land degradation on agricultural lands. The financial and administrative skills that landcare groups have, often gained through the receipt of previous government-provided grants, are crucial to the effective management of such devolved grant schemes. Furthermore, having a community based group of local individuals ensures a local commitment to the outcomes of the project; it is in the best interest of the committee that the programme works well because their local community will be the major beneficiary of the project. While an effective voluntary committee is necessary for the project, it is often advantageous for a dedicated project officer to be employed and take responsibility for the day-to-day running and administration of the project.

Farmers usually need more than just financial incentives to adopt new farming practices. This is particularly the case where the new practice provides limited direct private benefit and far greater public benefits. Through the on-going activities and projects of community landcare, rural communities have become sensitized to land degradation. Such groups have been effective in gradually changing the social norms surrounding farming practices, and thus more sustainable farming practices have become acceptable and more generally promoted.

## Lessons Learned

The increasing reliance of the government on landcare to achieve sustainable production outcomes has been criticised by some commentators as devolving state responsibility for conventional extension in agriculture. It is true that during the 1990s the number of state agency extension staff providing direct advice and extension to farmers decreased considerably. At the same time there has been a move away from solely production- and industry-focused extension to greater emphasis on public good extension, particularly related to land degradation. The delivery format has changed from a one-on-one relationship between farmer and extension officer towards group-based approaches. Group extension, as characterized by the landcare group, as a method of extension has been increasingly supported by professional extension workers because of a recognition of the value of farmer knowledge and a belief that groups are an effective means of facilitating the transfer of knowledge. Landcare groups are an important means of building social support for the adoption of more sustainable farming practices and of encouraging social norms more conducive to conservation among landcare's rural constituency.

The move away from using public monies to support private-benefit extension to that which delivers public-benefit is more likely to be supported by the non-rural Australian public. However, there is little incentive for farmers to invest in public-benefit work if they do not receive any return on that investment, or if the measures being advocated are complex or impractical at a farm scale. Furthermore, many "sustainable" farming practices will have both public and private benefit characteristics. The cost-sharing arrangements described for the Tod River Catchment suggest that farmers are willing to contribute to the cost of such activities even though they delivered some public benefit in addition to the private benefit. However, the types of activities readily taken up by landholders were those that appeared to deliver the greatest short-term gains in productivity, that is, provided the highest returns. For other activities, for example the conservation of biodiversity, the return to landholders is considerably reduced and this will be likely to reduce their willingness to participate, though clearly it will not prevent farmers from performing such activities.

## Contacts

The Landcare Contact Officer, Natural Heritage Trust Administration Section, Natural Resource Management Policy Division, Agriculture, Fisheries and Forestry - Australia, GPO Box 858, Canberra, ACT 2601. Telephone: (02) 6272 5474; Facsimile: (02) 6272 5618.

Sustainable Resources – Waite, Primary Industries South Australia, GPO Box 1671, Adelaide, SA 5001. Telephone: (08) 8303 9642; Facsimile: (08) 8303 9555.

Social Sciences Centre, Bureau of Rural Sciences, PO Box E11, Kingston, ACT 2604. Telephone: (02) 6272 3582; Facsimile: (02) 6272 4896.

# Chapter 7

# Madagascar: Contracting for Community Forest Management

Regina Laub-Fischer

In Madagascar numerous generations of rural extension workers busily taught and demonstrated better agricultural techniques to the rural population, explaining the destructive consequences of burning and destroying the forests and savannahs. Most of the donor agencies as well as the concerned Ministries spent huge amounts of money to improve the conditions for sustainable natural resource management. What is the result of these efforts?

In the highlands of Madagascar the landscape is covered with smoke from enormous bush fires; every year hundreds of square kilometers of natural forests are cut down and destroyed; and abandoned old terraces give evidence of the failed efforts of the rural extension staff. Previously the Ministry of Water and Forest (MEF) was responsible for the management and protection of the forests, although it did not have enough staff and lacked control over the forestry resources. The forests were exploited illegally by the local population but especially by concessionaires, as well as by the Ministry staff – with hardly any restriction. The forest was looked upon as common property for which nobody felt responsible. These natural resources were exploited as far as possible, but nobody thought about conservation or management of the forest. According to the statistics, Madagascar loses about 200 km² of forest per year. At present only 20% of Madagascar's surface is still covered with woods.

What are the causes for the attitudes behind such destruction? Are the generations of rural extension workers unable to pass the message on to the rural population, or are the generations of farmers too ignorant to understand their message? Or is it a question of unfavorable conditions?

Is there light at the end of the tunnel? During the last 4 years a contractual approach to assure the sustainable management of natural resources was developed and introduced by the integrated forestry project (PDFIV) which is implemented by the MEF, Ministry of Water and Forest, and the GTZ (Deutsche Gesellschaft fuer Technische Zusammenarbeit).

The main objective of the project is to develop agricultural and forest sustainability models together with the population, with a view to enabling them to manage their natural resources in a sustainable and economically beneficial way. Based on some encouraging, but rare instances where the rural population managed their own natural resources in a sustainable way, the Malagasy government took the decision to pass in 1997 the new forestry law and the concept of GELOSE (secured local management of natural resources).

According to this new forestry law the rural population are given authority to manage the forests and natural resources in their area by themselves, if they respect a certain management plan accepted by the Ministry. That means that the government has given back certain responsibilities to the local population that government has never been able to realize.

In line with these new conditions, the Ministry of Water and Forest and the local population collaborated in an extensive series of meetings, training and information sessions to develop a contract. The contract between the MEF and the local population was to be implemented by an NGO, the UFA (Union Forestière d'Ambatolampy or Forest User Association).

Since the signing and implementation of the contract, the situation has changed considerably. The local population feels more responsible for their forests as a result of the UFA management plans developed together with the population – for about 5,000 ha of forest.

According to these plans the rural population is determined, willing and able to manage the forests. As a result of this responsibility placed with the people, the forests are now well protected and controlled. An immediate effect is that the concessionaires no longer have free access to the forests, and control by the people is getting increasingly efficient.

The roles and responsibilities of the UFA, the population, and the MEF are clearly defined in the contract. The forest user association (UFA) employs forest technicians, a manager and some administrative staff and is responsible for the management of the forests in the District. It provides the technical know-how to the user groups and supports them in elaborating their management plans. Another UFA responsibility is the maintenance of the roads in the forestry stations and the marketing of the timber. When the UFA sells timber, then the Ministry receives a certain percentage of the benefits.

The Ministry of Water and Forest (MEF) is represented by a District representative for the supervision of the legal situation in the forest and the realization of the land use plans and the selling of the timber. The Ministry's supervision represents public concern for conservation of the most precious parts of the natural forests and ensures sustainable utilization of the forest resources.

The local forest user groups, i.e. the population along the forest border areas are organized into small forest user groups. A land and forest management plan is developed and presented by them for acceptance by the Ministry of Water and Forest. These groups have the responsibility to follow and implement the management plan. The user groups have several benefits: they can be hired and paid by the forest association to work in the forest. In addition to that economical benefit the group members have clearly identified areas in the forest where they can procure fuel wood, timber and other forestry products.


## Impact of the Contract

After long preparation, 1994-1998, the contract between the forest user association and the MEF was signed in 1998. During the last 1.5 years some changes in the management of the forests can already be observed.

**1.** The local population feels increasingly responsible for the forest. They protect it against illegal deforesting and respect the forest management plans. Uncontrolled exploitation has been stopped to a large extent.
**2.** The number of forest offenses has decreased considerably.
**3.** The forest fires have decreased significantly in the area of the project. In 1996, for various reasons, about 150 ha of the forest were burned on purpose by the local population. But in 1999 there was just one incidence – a fire broke out by accident destroying about 4 ha of forest.
**4.** The UFA management is efficient, as is their technical support for the population. For a few months the UFA has been completely financially independent. The forest user association can now manage about 5,000 ha of forest and thereby finance the management unit, maintain the roads in the forests, pay the workers, purchase material, etc.
**5.** The UFA has identified several clients who want to buy (or have already bought) timber in large quantities. The marketing of the timber is not yet well organized however and requires much more attention and supervision from the UFA.

Besides these general observations some major benefits for the different groups which are involved in the contract are also noticeable.

**1.** While working in the forest and cleaning certain parcels, the forest workers discovered a new income generating scheme, that the little straight trees, which have to be cut down during the thinning of the forest, can be prepared and sold as broomsticks.
**2.** Some women's groups have established tree nurseries to produce local tree seedlings to diversify some parcels of impoverished forest.
**3.** The rural population has permission to collect wood after the thinning of the forest. They use it for charcoal production and manage to satisfy about 50% of the quantity of charcoal needed for the main town of the District.
**4.** The target for the UFA is to sell 3,000 $m^3$ of timber per year. This target has not been achieved so far due to technical problems of storing, sawing and selling the timber. However, local entrepreneurs plan to construct a new sawmill in the

District. During the past year these technical problems are beginning to be solved and production is expanding.

**5.** Employment has increased in recent years in the District due to the prospering timber production. An increasing number of entrepreneurs interested in timber processing have settled in the main town of the District.

**6.** The representatives of the Ministry limit their activities to supervising the UFA and intervene only rarely, for instance, if there are forest offences.

**7.** The local population directly benefits from their own efforts; they have an immediate economic benefit from working in the forest. Furthermore they can make the decision as to whether someone can cut down a few trees in their forest, instead of walking 50 km to deposit their demand at the local forest administration office.

The following four critical success factors were necessary to achieve the benefits experienced by the project team and the population:

**1.** The legislation and the changing forest law made it possible for the approach of the community-based forest management to be undertaken on a legal basis. These changing conditions are also one of the main reasons why the approach was copied in other areas of Madagascar.

**2.** The establishment of effective communication mechanisms between the concerned groups and organizations was also essential. In order to achieve good collaboration between the different entities, the following steps were most important:

- People in the villages, but also in the Ministry had to change their mentality from unlimited exploitation to sustainable management.
- The local population had to be engaged in this process. The project made an effort to organize people into small forest user groups working together under the supervision of the UFA. This structuring process and organization of the rural population is a key element of the functioning of the UFA.
- Changing the role of the Ministry employees was difficult because these officials were afraid to lose power and influence, as well as the benefits from timber.
- In order to implement the contract, the rural population had to be informed and involved in the whole process. For example it took much time to discuss with all the concerned parties the different aspects of the contract and to find a consensus among all parties involved.

**3.** Another crucial factor was the elaboration of the forest management plans together with the local population. The wishes and needs of the population were considered and combined with the needs of good forest management. This participation of the people during the planning process led us to a situation in which the people were willing to work in the forest according to the plan.

**4.** Finally, another crucial factor was the marketing of the timber by the UFA. During the first stage of the contract, the managing unit of the UFA had to learn how to handle the stocking, transportation and selling of the timber.

The forest management plans were not limited just to the forest areas. The approach covered the whole territory with all its natural resources. Due to this approach all the people in the territory were involved in the planning and implementation – including women, the young and the old.

## Sustainability and Replicability of the Approach

Contracting approaches promoting community management of forests is spreading in numerous areas in Madagascar.

It appears that a certain "win-win situation" has been found: the local population earns money while working in the forests, the forest products improve in their quality, and the UFA can sell the timber for a better price. The MEF can clearly see the difference in the actual situation as compared with that prior to the UFA's management and technical support, specifically the decreasing number of offenses and fires destroying the forests. As long as the UFA and the contractual relationship work, the economical and ecological benefits for the rural population are assured.

Additionally, the plans are to be adjusted every 5 years. While elaborating the plans with the local population, different zones are to be identified, taking into consideration the population's special needs and problems. If necessary, people can also change their plans with the approval of the Ministry.

In order to upscale the contracting, work has to be done - especially by the Ministry for Water and Forest. So far the areas where contracting has occurred are still considered "pilot areas" – areas of trial and outside the normal context. One additional possibility would be to increase the price for timber on the national and international market considerably.

## Lessons Learned

In general it was difficult to convince the MEF of the importance of such a contracting relationship. The project worked hard for about 4 years to convince the MEF of the utility of this contract.

Full participation of the people at all the levels of the process was essential to strengthen the rural population, to take them serious and to involve them in all the different steps of the negotiations with the Ministry. The rural population was supported by forming small forest user groups who were willing and able to manage a few parcels of the forest. These basic group formations were strengthened by regulations regarding the utilization and management of the forest. Also, the social behaviour of the group members was precisely identified.

Due to the rather limited presence and influence of the Ministry in the rural areas, the population has had to take over even the supervising role of the

Ministry. The mutual control mechanisms of the different small forest user groups have become very efficient and accepted by the people as long as the decisions are based on the forest management plans.

All these efforts would have been more or less in vain without the change in political conditions. Extension managers or international staff who are interested in learning more about these tools can get support and help from the integrated forestry project. Training modules are available for study. The project has already prepared visiting programmes to demonstrate and explain in detail the contracting approach to community forest management.


## Contacts

M. Mussard Anrdiamanana, the National Director of the Project, and
M. Mosa Ranaivomanana, the person in the project responsible for sustainable forest
    management. (Both of them followed closely the different steps of the UFA and its
    development.) Both can be contacted at:
PDFIV, B.P. 869, Antananarivo 101, Madagascar.
Email: pdfivamb@dts.mg

David Andrianjafy, Manager of the UFA
PDFIV, B.P. 42, Ambatolampy, Madagascar

M. Philibert Rarivomanana, Directeur de Programmation et de Suivi et Evaluation
Ministère des Eaux et Forêts
B.P. 243, Antananarivo 101, Madagascar

M. Rivo Ratsimbarison, Chef de Cellule Gelose and
M. Haribenja Andriantsoavina
ONE (Office National de l'Environnement)
B.P. 822, Antanananrivo 101, Madagascar

# Section Three

# CONTRACTING FOR INPUT SERVICES

# Chapter 8

# Bangladesh: Sub-contracting Extension Services to a Local Private Agricultural Training Institute

M. Hassanullah*

## Location of the Case

This case occurred in Sherpur thana of Bogra district in the northern region of Bangladesh. The main partners in the contract are M/s Mokbul Hossain, Mahipur Agricultural Training Institute (MATI) and Agrobased Industries and Technology Development Project (ATDP).

Mr. Md. Mokbul Hossain became a seed dealer in 1985 and performed well in his seed business. He acquired a fertilizer dealership from Bangladesh Chemical Industries Corporation (BCIC) in 1989 and became a member of the Chamber of Commerce and Industry in 1996. Mr Hossain was impressed with the performance of hybrid maize demonstrated by Fertilizer Distribution Improvement (FDI) II project in 1990. He became an agent, selling hybrid maize seed for a national seed company. He then promoted the cultivation of maize through a contact growing system, to expand his seed business as well as to trade grain with traders, retailers and poultry farm owners.

MATI is a private institute managed by Karatoa Multipurpose Cooperative Society Ltd. The institute offers 3-year agricultural diploma courses recognized by the Bangladesh Technical Education Board. It has a well-qualified and experienced teaching staff that offers courses on crop production, animal and fish farming and related disciplines of soils, agri-machinery, agro-processing, etc.

---

*** Note:** This is one of the cases selected by the editors from among three cases of sub-contracting agricultural extension services in Bangladesh provided by Dr M. Hassanullah.

## Country Context

Some indigenous maize varieties are traditionally cultivated in Bangladesh by tribal and ethnic communities who migrated from India. As the animal industry began to grow in the early eighties, demand for maize increased to feed animals and poultry. Much of the demand was being fulfilled by imported maize. In order to increase local production of maize the Bangladesh Agriculture Research Institute (BARI) developed open pollinated high yielding varieties of maize and introduced their cultivation in different regions through the Department of Agricultural Extension (DAE). The yield potential of these varieties of maize was not attractive enough to be widely disseminated. Nevertheless, the demand for maize continued to grow and the high price of feed was deterring the growth of the country's animal industries.

In this context the private seed companies introduced the hybrid maize. The DAE and some non-governmental organizations (NGOs) also began to demonstrate the cultivation of hybrid maize. Land in the Sherpur region is most suitable for maize in the winter crop season because after the recession of floodwaters, marginal crops are cultivated or the land remains fallow for pre-monsoon planting of deep-water paddy. Mr Hossain foresaw the prospect of expanding grain trading manyfold by promoting cultivation of maize in the region. He began to promote the cultivation of hybrid maize through a contract growing system.

In the winter of 1995/96 ATDP showed a documentary on hybrid maize cultivation using balanced manuring and fertilization which inspired many farmers to cultivate hybrid maize. Mr Hossain approached the ATDP for assistance to provide extension services to the farmers in his region. Most of them lacked the knowledge and skills of cultivating hybrid maize necessary to exploit the full production potential of maize. Eventually, Mr Hossain, on behalf of his contract farmers, received technical assistance to educate about 500 farmers. ATDP granted the assistance and engaged MATI, as recommended by Mr Hossain, to provide extension services to a large number of farmers on a contractual basis.

## Description of the Contract

ATDP entered into a grant contract with M/s Mokbul Hossain to provide extension services to 480 contract farmers to master the technologies of hybrid maize production, protection and post harvest handling and advise M/s Hossain on marketing of maize. Following the grant agreement M/s Mokbul Hossain entered into a contract with MATI to assist in: (i) selecting proper types and varieties of maize to produce; (ii) helping Hossain's 480 contract growers to master the technologies of maize production, protection and post harvest handling through training and post training supervision, guidance and counselling; and (iii) advising Mr Hossain on guidelines for profitable marketing of maize. ATDP reserved the authority to pay MATI on issuance of a certificate of accomplishment by M/s Mokbul Hossain.

## Impact

Training and post training advisory support by MATI to the contract farmers of M/s Mokbul Hossain have created a significant impact on the production and trading of maize for M/s Mokbul Hossain in particular and the entire region at large. Production and trading trends are shown in Table 8.1.

The number of contract farmers, production and sales of maize have increased by over 100%. As a result, area and production of maize have increased manyfold in the whole region (Table 8.2).

**Table 8.1.** Production and trading of maize by M/s Mokbul Hossain.

|  | Year of production | | |
| --- | --- | --- | --- |
|  | 1995/96 | 1996/97 | 1997/98 |
| Number of contact growers | 153 | 203 | 482 |
| Land covered through contact growing (ha) | 130 | 350 | 700 |
| Amount of seeds sold (t) | 2.644 | 7.0 | 14.0 |
| Maize purchased and sold (t) | 215 | 337 | 843 |

**Table 8.2.** Area and production of maize in Sherpur Thana.

|  | Year of production | | | |
| --- | --- | --- | --- | --- |
|  | 1994/95 | 1995/96 | 1996/97 | 1997/98 |
| Land covered by hybrid maize (ha) | 118 | 148 | 445 | 1,216 |
| Production (t) | 708 | 962 | 3,100 | 7,464 |
| Price value (Taka) | 5,310,000 | 7,215,000 | 24,025,000 | 37,340,000 |

## Sustainability and Replicability

The public sector agricultural extension services are unprepared to provide need-based intensive extension services to private traders and a large number of contract farmers with the latest technologies of production, protection and handling of a particular industrially important crop like maize. NGOs also lack technical competence to provide such intensive services for dissemination of high technologies.

This service contract met a well-defined need that could be provided by a well-established private agricultural institute like MATI having physical facilities as well as technical expertise. Only recently have a few such private agricultural institutes and colleges emerged in Bangladesh in response to the current privatization policy of the government and the increasing demand for agricultural education. This contractual system is seen to be sustainable because the trader, Mr Mokbul Hossain, and the farmers benefit equally.

Evidence shows that the production and trade of maize have increased considerably. But further follow-up services are needed to educate the new farmers emerging each year, as they may not have access to the knowledge and skills acquired by the experienced farmers. As a result of the success of the project, the ATDP has been replicating similar sub-contracting arrangements in several other regions for production and trade of industrially important field crops such as maize, soybean, sunflower, sorghum, cassava and export potato.

## Lessons Learned

This case highlights an arrangement whereby the specialized educational needs of a specific and targeted extension clientele are provided by a local private agricultural training institute. The case demonstrates how local private agricultural training institutes and colleges can be utilized effectively to provide short and intensive training of large numbers of farmers in a particular skill and provide follow-up advisory services for practical use of technologies learned. This underscores the social role of such private institutions, which can reduce the responsibility of the public sector extension organizations. Such arrangements can be effective when the tasks are specific and the recipients have control over service providers. Additionally, the case indicates that extension managers should set aside a part of their operational fund for such sub-contracting when specific demands of specialized services arise. In short, the case implies that extension managers should seek to identify local institutions to fulfill specific training needs instead of trying to build such specific facilities within the extension system that would be utilized only occasionally.

## Contacts

Dr Ronald P. Black, Chief of Party, ATDP, IFDC, House 103, Road 1, Block F, Banani, Dhaka 1213, Bangladesh
Fax: 880-2-988 1724
Email: atdp@citechco.net

Principal, Mahipur Agriculture Training Institute (MATI), Mahipur, Sherpur, Bogra, Bangladesh

Mr Mokbul Hossain, Proprietor: Tusher Beezaghar, Dhanut Road, Sherpur, Bogra, Bangladesh

Chapter 9

# Mali: Contracting for Livestock Production Extension with Private Veterinarians

E. Fermet-Quinet and J. Gauthier

## Introduction

One of the growth areas for extension in the livestock sub-sector is the emergence of the private veterinary sector in providing animal health services, including extension services. Following their structural adjustment programmes, governments and donors have favoured the establishment of a network of private veterinarians in most of the developing countries in an effort to reduce public costs. In Africa, for instance, more than 2,000 veterinarians are currently practicing privately. Even in state-dominated China, livestock complexes such as feedlots, pig and poultry farms have contracted over 3,000 private veterinarians or paraveterinary staff, many of whom have transferred from government service.

This private animal health network provides a good opportunity to improve livestock production extension. On the one hand, private veterinarians like to diversify their activities and increase their clientele and income. On the other hand, governments need to support the private sector and reduce the costs of their extension services. By paying private agents to provide extension services they achieve both objectives.

## Identification of the Case

Mali has been very innovative in developing its private veterinary network and involving it in extension work, especially through a project called PASPE (Projet d'Appui au Secteur Privé de l'Elevage) funded by the French Co-operation. Since 1998, the project has supported the livestock development of three regions of Mali, which are Sikasso, Kayes and Mopti. The innovations are: (i) the use of a network of private veterinarians and paraveterinarians to transfer information on livestock; and (ii) the participation of the producers through Producer Organizations (POs) and Regional Agriculture Chambers (RACs) in extension programming and contracting. An international non-governmental organization (NGO), VSF (Vétérinaires Sans Frontières), was involved in organizing village consultations for identifying extension activities. The supporting NGO was also in charge of training the future trainers (i.e. the private veterinarians) and supporting RACs in contracting.

## Country Context

In the three regions mentioned above, livestock production extension (LPE) was carried out by the Ministry of Rural Development and Water (MoRDW). The LPE messages were the same for the three regions and considered too theoretical and often inappropriate to the local demand. The extension programme was also considered too costly with very low impact on livestock productivity.

However, the context has changed since 1991. Mali has been engaged in privatizing, decentralizing and restructuring its agricultural services. Through a strong policy of privatization, Mali has favored the development of private veterinary units and 213 private practices were established in 1998. In the context of decentralization, RACs were established to represent POs at the regional and national levels. Finally, following the restructuring of the public sector, extension staff of the MoDRW was reduced leaving opportunities for alternatives and limiting unfair competition with the private sector.

## Types of Contracts

In 1998, PASPE started a programme for supporting livestock development in Sikasso, Kayes and Mopti. In order to strengthen the private sector, reduce the operating costs of the support services and promote a more demand-driven extension system, the project proposed contracting private veterinarians or paraveterinarians through the RACs after having selected regional extension programmes through consultations with the producers at the village level.

From village consultations eight extension themes covering poultry, small ruminant and cattle production and animal health were identified in 1999. The RACs contracted 120 private agents for providing training in 4,500 villages (two visits per year). The private agents were paid on the basis of the number of villages visited. The costs were only US$20 per village and per year. Extension messages were provided free to farmers. Vaccines, drugs and other material such as fly traps had been charged at full cost by the private veterinarians.

A contract was established between each private veterinarian and the RAC. The veterinarians or paraveterinarians were due to report the number of training sessions organized. The RAC was in charge of paying the private agents after having confirmed that the work had been done. The control was based on a survey carried out in a sample made up of 15% of the villages reached by the private extension agents. One false record was enough to cancel the whole due payment and contract.

At the end of 1999, following the first round of training sessions organized at the regional level, the demand from the herder organizations was getting increasingly diversified between villages of the region of Sikasso. Consequently, the PASPE sought more flexibility in selecting the extension themes and launched the village extension vouchers (chèques formation) provided through the RAC of Sikasso.

The RAC has extension funds (vouchers) available for each village (US$17 per village) and edits a catalogue presenting the training options, a list of authorized trainers and the costs for each training. The villages are in charge of selecting the training sessions they need. They have to participate in funding if training costs exceed their extension allocation. Farmers can also ask for other training options than those presented in the official list. The new themes have to be agreed upon by the RAC and organized in collaboration with the supporting NGO. Villages can combine their allocated funds if necessary. The villages select the trainers themselves. If a trainer is based far from a village, additional transport costs are paid partly by the village. A contract is signed by the representative of the village, the private agent and the RAC.

In addition to more flexibility for selecting extension themes, the village extension vouchers improved the empowerment of the producers, transferred part of the control to the beneficiaries and allowed for competition between extension agents.


## Impact

The major beneficiaries of the new extension system are the livestock producers. Access to support services and inputs has been improved in the regions covered by the project. For the year 1999, the results are the most impressive in the region of Sikasso where about 40 private agents were involved:

**1.** 1,850 villages have received advice on poultry diseases and 2,500,000 poultry have been vaccinated against Newcastle disease.
**2.** 1,250 villages have received information on sheep and goat diseases and 15,000 small ruminants have been vaccinated.
**3.** 175 villages have attended training in trypanosomiasis control and 500 fly traps have been sold.
**4.** 425 villages have been trained in range and water management.

The other beneficiaries of the project are obviously the private veterinarians and paraveterinarians. They have diversified their activities and increased their income. Each of the private agents earned an average US$740 from the RACs.

producers of their zone and have therefore strengthened their clientele and increased the sales of drugs and vaccines.

The final beneficiary is the government. The cost of such an extension programme is very low and the network of the RACs is reinforced, improving the dialogue between the MoRDW and the POs.

The impact was assessed by measuring the number of villages reached and the numbers of farmers trained and by estimating the inputs sold. No data related to the agricultural productivity are available.

Four critical success factors follow:

**1.** Strong participation of the herders in selecting priorities.
**2.** Development of extension programmes with rapid and tangible impact such as a decrease of mortality rates.
**3.** Extension messages associated with improving access to commercial inputs (drugs, vaccines, traps, feed, material).
**4.** Individual contracts for each private agent for insuring appropriate logistics, refresher training and strict control.

## Sustainability and Replicability

The involvement of the POs and the private sector are good criteria for sustainability.

The costs paid by a village to the private extension agents are about US$20 per year. The low costs are also a strong advantage in favour of the sustainability of the system. However, there is a need to explore the willingness of the producers to participate in paying private agent services in addition to inputs.

The overall annual costs, including training of the private veterinarians, consulting POs, organizing the annual programme and controlling the work done by the private agents are about US$320 per village. Having contracted an international NGO is the main reason of the high costs. Therefore, there is a need to explore the opportunities of transferring support activities to local NGOs or to technical units based in the RACs.

Replicability is feasible provided that: (i) the network of private agents already exists and just needs to be strengthened; and (ii) decentralized institutions are functional.

## Lessons Learned

Although the system is based on contracting the private sector through the RACs who represent the producers, there is a need to involve the public sector so as to prevent unfair competition and misinformation. The public sector has a role to play in supporting the training of the private agents. There is a need for continuous refresher training in order to assure extension quality. Also, lack of transport is often a major constraint when contracting a private agent. Therefore, providing private agents with access to credit in order to allow them to buy motorcycles is in this case a key element of success.

# Chapter 10

# USA-Illinois: Contracting for Precision Agricultural Services

Burton E. Swanson, Mohamed Samy and Patrick D. O'Rourke

## The Case and the Setting: Precision Agricultural Services in Illinois

The agricultural technology transfer system in the state of Illinois has been progressively privatized during the past 40 years as input supply firms, including many farm cooperatives, incrementally hired Certified Crop Advisers (CCAs) to provide technical advice to farmers in the areas of crop production, soil nutrient management, soil and water management, and integrated pest management. Certified Crop Advisers are members of the American Society of Agronomy (ASA) who have met the requisite educational qualifications and passed both international and state-level qualifying examinations. In addition, CCAs are required to obtain 40 hours of in-service education every 2 years and agree to a code of ethics.

During this period, the number of public agricultural extension workers declined from about 100 farm advisers for the entire state in 1960, to approximately 20 cropping system educators by 2000. By the end of the 1990s, the number of CCAs employed by private sector firms and farmer cooperatives, or working as private consultants, had increased to more than 1,000 advisers or a ratio of about 1 CCA to 70 full or part-time farmers.

Precision agriculture involves the use of geographic information system (GIS) software to allow the precise mapping of a field. Once mapped, soil samples are taken on the basis of 1, 1.3, or 2-ha grids. After these soil samples are tested, the results are fed into a computer programme that generates maps showing the current level of soil fertility or nutrient deficiency for phosphorus (P), potassium (K), nitrogen (N), and soil acidity (pH) throughout each hectare grid within the field. This process is referred to as *grid soil testing*.

The results of grid soil testing are used to generate a digital set of fertilizer (P, K, and N) and lime recommendations, again for each hectare within the field.

This analysis produces a *soil fertility map* for each major element, plus pH. Next, variable rate technology (VRT) application equipment is used to apply the precise amount of each nutrient across each hectare of the farmer's field. Another precision agricultural tool being adopted by some farmers is the "on-the-go" *yield monitor*. This monitor is mounted on the farmer's combine to measure and record grain yields from each hectare of the farmer's field. Data from this yield monitor can then be used to generate a *yield map* showing productivity differences for each hectare across the entire field or farm. Also, these yield maps can be used to identify drainage problems and to compare yield difference between different corn hybrids or soybean varieties. Finally, this same yield information can be used to determine the amount of nutrients being removed from each hectare in planning the next year's soil fertility programme.

In 1990, a farm cooperative in a five-county area of eastern Illinois introduced precision technology into Illinois. By the mid-1990s, the use of precision technology for P, K, and pH had spread to many input supply firms within the state. Only once it had been commercially demonstrated that precision technology produced measurable economic benefits and that farmers were willing to pay for these new services, did this technology start becoming more widely available within the state, especially since 1995. Even today, 10 years after its first introduction, precision agriculture services are still not available to some Illinois farmers, especially in the poorer areas of the state.

## Contracting for Precision Agriculture Services

Contracting for precision agriculture services is between the individual farmer and either the local farmer cooperative, a private input supply firm, or a private agricultural consulting firm. Under each arrangement, farmers generally contract for one, two, or three possible levels of service, including: (i) grid soil testing; (ii) generating a digital soil fertility map with site specific recommendations; and (iii) VRT that involves the actual application of different production inputs, generally using custom applicator equipment, at the recommended input levels across each hectare of a field. It should be noted that VRT for nitrogen (N) has been commercially tested in selected areas of the state since the mid-1990s to reduce nitrate/nitrite runoff, thereby offering a potential solution to important environmental problems associated with the high use of nitrogen fertilizer. This case will be limited to the adoption and use of precision technology for P, K, and pH applications.

Grid soil sampling and testing costs about US$3 per acre or US$7.50 per ha. The farmer receives the results that show soil nutrient deficiencies throughout each 1-2 ha grid within his/her field. This grid soil sampling and testing service is generally carried out on a 4-year cycle or rotation, so that the farmer tests one-quarter of his/her farm each year to spread out these costs. On the basis of this information, the farmer can determine whether the soil acidity (pH) and fertility is sufficiently heterogeneous within each field to justify the additional expenditure of generating a set of digital soil fertility maps that can be used for VRT application. If so, the service provider will generate N, P, K, and lime recommendations for each grid area within the farmer's field. This second level of service generally costs an additional US$7.50 per ha. These soil fertility maps can be fed into the computer on the VRT truck at the time of custom application.

The third level of service is applying different production inputs within the farmer's field using VRT equipment. The average cost of each VRT application is US$5 per acre or US$12.50 per ha. It should be noted that custom fertilizer application, using a blanket application rate, averages US$3 per acre or US$7.50 per ha, therefore the additional cost of VRT is about US$2 per acre or US$5 per ha. In addition to purchasing these different services under contract, many Illinois farmers have purchased GPS yield monitors so that the farmer can measure and digitally record the yield or output of grain from each hectare within his/her field. As noted earlier, this computer-generated information is used to produce a yield map to show production levels and nutrient removal for each grid area within each field. Figure 10.1 illustrates fertility and yield maps for an Illinois farm, including recommendations for either maintaining, or maintaining and building the soil fertility.

The adoption of precision technology generally occurs incrementally in terms of the specific tools to be employed (i.e. grid soil testing, soil fertility maps, VRT, yield monitors, and so forth). In addition, these services are generally phased in over time. For example, since grid soil testing is recommended every 4 years, most farmers start by testing one-quarter of their field in year one, the next one-quarter in year two, and so forth. This approach allows the farmer to test the efficacy of this new technology on part of his/her farm before applying these tools throughout the entire farm. This approach also allows the farmer to get the farm on a 4-year rotation, thereby, spreading these soil-sampling costs across 4 years. Other costs, such as the use of VRT, are annual costs, while the purchase of a yield monitor and computer are capital investments.

## Impact of Precision Agriculture

### Adoption of Precision Technology by Illinois Farmers

In studying the impact of precision agriculture within Illinois, data were collected in two areas of the state, but outside of the original area of introduction. The first target area includes six relatively heterogeneous counties in northern Illinois and involved 770 respondents out of a total population of 3,113 farmers or a response rate of 24.7%. These data were collected between August and early October 1999 for the 1999 growing season.

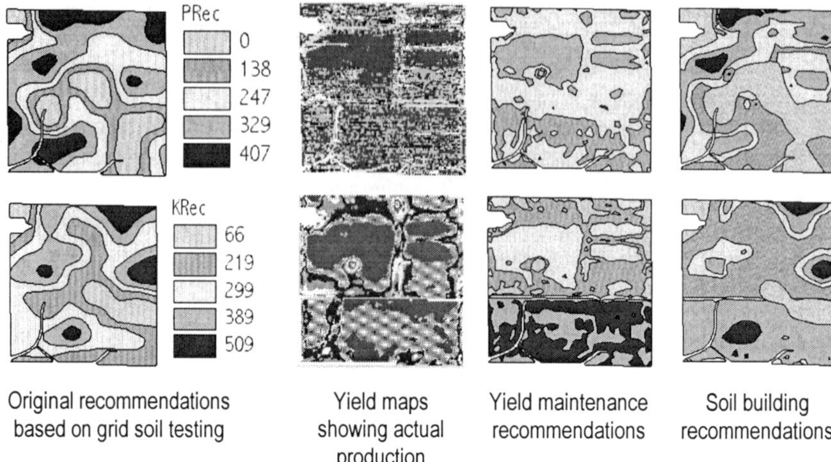



Original recommendations based on grid soil testing

Yield maps showing actual production

Yield maintenance recommendations

Soil building recommendations

**Figure 10.1.** Illustrative fertility (P & K) and yield maps for a 64-ha Illinois farm.

The second study area involves McLean County, a large, relatively homogeneous county in central Illinois. In this case, survey data were collected during two time periods, with a specific focus on the adoption of precision agriculture technology. Data collection for the first survey was carried out during March-April 1998 for the 1997 growing season. Then, this survey was repeated during November-December 1999 for the 1999 growing season. The data set for the 1997 growing season involved 542 usable questionnaires out of a total population of 1,441 potential farmers or a 37.6% response rate. The 1999 data set involved 343 usable respondent questionnaires out of a total population of 1,407 farmers or a 24.4% response rate. In both study areas precision agriculture technology first became commercially available to farmers in 1995.

In the first study area in northern Illinois, which illustrates the effects of socioeconomic factors on the rate of farmer adoption, it was found that one-third of the farmers had adopted at least one precision technology component by 1999, including: 33.6% using grid soil testing, 27% using soil fertility maps with site-specific recommendations, and 31.4% using variable rate technology for the application of lime and/or fertilizer. In addition, another 35% were considering the adoption of one or more precision technology components or tools.

Adoption of precision technology was significantly related to farm size, age, and educational level. For example, 57.5% of farmers with more than 1,000 acres were currently using grid soil testing and VRT. The percentage of adoption of grid soil testing was 35.3% for farmers with between 500 and 999 acres, and only 23.3% for farmers with < 500 acres. The adoption of VRT followed a similar pattern, with 31.3% of farmers with between 500 and 999 acres and 15.4% of farmers with < 500 acres adopting this third level of service.

The farmer's educational level was also related to the adoption of precision farming technology. For example, 42% of the farmers with 16 years of education (i.e. a BSc degree) had adopted grid soil testing. For the same component, only 32.9% of farmers with 13-15 years and 28.8% of farmers with 12 years or less education had adopted this component. A similar, but less differentiated, relationship was found between educational level and VRT.

In determining the potential impact of farmer's age on the adoption of grid soil testing, the respondents were divided into two age groups: 50 or more years, or less than 50 years of age. For grid soil testing, it was found that 37.5% of the farmers who were less than 50 years of age had adopted this component, in contrast with 29.9% for farmers who were 50 years or older. The same pattern was detected for the use of VRT. The percentage of farmers who are considering the adoption of these technologies also favours the younger farmers.

The second study examined the change in adoption of different precision technology tools. As shown in Table 10.1, 22% of the total respondent group had adopted grid soil testing with GPS by 1997. Of this adopter group, farm size was very important, with 46.3% of respondents with 1,000 or more acres adopting this component, 22.6% of respondents with 500-999 acres adopting, and only 12.5% of respondents with < 500 acres adopting. The same pattern was detected in 1999; the percentage of large-scale farmers using this precision technology had declined, while the percentage of smaller- and medium-scale farmers using grid soil testing had increased. Consequently, the total adoption rate of grid soil testing had only increased to 23.5% by 1999.

**Table 10.1.** Adoption rate or percentage of farmers contracting for different precision technology services by size of farm in McLean County, Illinois.

| Size of farm (acres) | Total no. operator respond | GPS | | Variable rate application | | | Use of yield | |
|---|---|---|---|---|---|---|---|---|
| | | Grid soil testing | Nutrient id map | Lime | P & K | N | Monitors | Maps |
| **1997** | | | | | | | | |
| < 500 | 288 (53%) | 12.5 | 4.9 | 13.5 | 10.8 | 4.9 | 8.3 | 5.6 |
| 500-999 | 146 (27%) | 22.6 | 11.0 | 22.6 | 11.0 | 8.2 | 13.0 | 8.2 |
| 1000+ | 108 (20%) | 46.3 | 24.1 | 37.0 | 22.2 | 15.7 | 42.6 | 29.6 |
| Total | 542 (100%) | 22.0 | 10.3 | 20.7 | 13.1 | 7.9 | 16.4 | 11.1 |
| **1999** | | | | | | | | |
| < 500 | 216 (49%) | 18.1 | 7.4 | 17.6 | 12.0 | 4.2 | 13.4 | 10.6 |
| 500-999 | 119 (27%) | 24.4 | 8.4 | 15.1 | 17.6 | 4.2 | 21.0 | 16.8 |
| 1000+ | 107 (24%) | 33.6 | 15.9 | 26.2 | 20.6 | 3.7 | 43.0 | 33.6 |
| Total | 442 (100%) | 23.5 | 9.7 | 19.0 | 15.6 | 4.1 | 22.6 | 17.9 |

As shown in Table 10.1, the adoption of VRT in applying lime closely parallels the adoption of grid soil testing. The reason why VRT lime application is relatively widespread is because soil pH levels vary considerably within individual fields; therefore, VRT for correcting soil pH saves farmers money. However, the adoption of VRT for P, K, and N was found to be considerably lower. The reason for this lower adoption level is that most farmers have tended to apply relatively high levels of P and K over many years. Therefore, fertility levels tend to be fairly high and uniform. Consequently, it is not economic for farmers to use VRT for N, P, and K in most fields, unless fertility levels are found to be highly variable.

The adoption of yield monitoring technology also increased between 1997 and 1999. Farm size was an important factor in the early adoption of this technology, with 42.6% of the large farmers adopting yield monitors by 1997. However, the primary increase in the use of yield monitors between 1997 and 1999 was among small- and medium-scale farmers. Finally, the adoption of yield mapping was also related to differences in farm size. A higher percentage of large-scale farmers had adopted this technology in 1997, but 2 years later this technology was also being taken up by small- and medium-scale farmers. Still, only about 20% of all farmers had adopted yield monitors and mapping by 1999.

The adoption of precision technology tools increased rapidly in McLean County between 1995 and 1997; however, there has not been any significant increase during the past 2 years. The apparent reason why this technology has not continued to spread is because corn and soybean prices have been at an all time low during the past 2 years. Therefore, most farmers have pursued cost reducing strategies, such as delaying machinery purchases and the adoption of precision technology services and tools. However, as commodity prices improve and farm sizes increase, it appears likely that farmers who manage heterogeneous fields and farms will adopt and continue using those precision technology tools that increase their productivity.

## Environmental Impacts

One important aspect of precision agriculture is its potential to reduce the negative environmental impacts of high input agriculture. For example, given the relatively low cost of nitrogen, Illinois farmers typically apply higher than recommended levels, with the goal of maximizing yields. However, this practice has resulted in surplus nitrogen being leached out of the soil each spring, primarily through the extensive system of field drainage tiles found throughout the state. Some of these excess nitrates or nitrites, depending on soils and other factors, end up either in freshwater reservoirs used for human consumption, or they flow downstream through streams and rivers and eventually end up in the Gulf of Mexico, contributing to the growing hypoxia problem offshore from the Mississippi River delta (a biological dead zone now equivalent in size to the state of New Jersey). Hypoxia is a serious problem affecting coastal estuaries worldwide.

As shown in Table 10.1, the adoption of VRT for nitrogen application is very low in McLean County, with the rate actually declining between 1997 and 1999. The reasons for this low level of use are twofold. First, the benefit of precision technology in applying nitrogen fertilizer on some soils is primarily a public good (e.g. reduction of nitrates/nitrites in public drinking water and reduced hypoxia), rather than a direct economic benefit to farmers. Second, even farmers who are concerned about these environmental impacts will be reluctant to utilize this technology unless there is a clear economic payoff or they are forced to reduce nitrogen use through the imposition of environmental regulations. Finally, it should be noted that similar environmental benefits could be achieved through the application of precision technology to pesticide application, especially those agrochemicals affected by soil texture.

## Sustainability, Replicability and Lessons Learned

Precision farming represents an important technological innovation that is spreading worldwide. The driving force behind this technology is that farmers can make site-specific management decisions, eventually involving seed, fertilizer, and agro-chemical recommendations that are both cost-effective and environmentally sustainable. Elements of this technology have potential application in developing countries to substantially improve the efficient and effective use of production inputs within specific agro-ecological zones. Precision farming will likely spread first to large commercial farmers in developing countries, probably on a contract basis, as is currently the case among commercial farmers in North America, Europe and Oceania.

Precision technology also has potential in high agricultural input countries with large numbers of small-scale farmers, such as China and India. These countries may determine that it is in their *national interest* to utilize selected precision agriculture tools – such as grid soil testing and soil fertility maps – to improve technical recommendations and, thereby, increasing productivity while reducing negative environmental impacts due to excessive input use. Under these situations, the grid area might be somewhat larger (e.g. 4, 9 or 25 ha grids), depending on the heterogeneity of soil conditions, water management practices, and farming systems. In addition, precision technology may serve as a potential substitute for well-trained agronomists who are unwilling to live in rural areas. Village level workers or farmer technicians could utilize fertility maps to make available accurate technical recommendations for farmers in different farming areas of each village.

## Recommended Sources on Precision Technology

Ciba Foundation Symposium (1997) *Precision Agriculture: Spatial and Temporal Variability of Environmental Quality.* John Wiley & Sons, New York.
National Research Council (1997) *Precision Technology in the 21st Century.* National Academy Press, Washington, DC.

## Section Four

# CONTRACTING FOR SPECIALIZED SERVICES

# Chapter 11

# Colombia: a Semi-private Coffee-growers Contracting System

Rodolfo Rodríguez and Tito Rodríguez

## Introduction

This case is about the technical assistance contracting arrangement entered into by two levels of government (municipal and national) and the provincial coffee growers committee, in Páramo municipality in Santander, Colombia. Originally, in 1992, the UMATA (Unidad Municipal de Asistencia Técnica Agropecuaria) of Páramo municipality worked by means of a contract with a private firm called SAGROVER. During 1993-1994, this firm became part of the structural organization of the municipality with exclusive co-financing by the DRI Fund (the Integrated Rural Development Investment Fund).

The UMATA is the Agricultural Technical Assistance Municipal (Local) Unit, created in all the municipalities by a National law in 1989. DRI is a centralized institution that finances and gives direction to the country's rural development policy.

In 1995, a new arrangement was initiated in Páramo municipality. This case study examines the new arrangement as it developed from 1995 to 1998. The following groups took part in this technical assistance (extension) contract system: the municipal administration, the provincial coffee growers' committee and the national government. The growers committee is part of the Coffee Growers National Federation, founded in 1927. It is a legal entity according to the Civil Law, which has the characteristics of a union and whose fundamental aim is the defense and the development of the national coffee industry.

The national government's economic assistance was provided through agreements with the municipality through the DRI Fund. Under this new contracting system, changes occurred in the administrative and operative management of the resources: in personnel selection methods, in working methods, in the control system, and in the coverage and monitoring of services offered to the users. These changes were basically due to the experience of the coffee growers' committee and the long trajectory (around 40 years) over which they viewed rural extension.

At present (February 2000) the extension service in Páramo is provided once again by the municipal administration, due to the budget reduction of the coffee growers' committee – resulting from the fall of coffee prices internationally. This circumstance limited the number of UMATAs co-financed under the contracting system.

## Contract Description

The objective of the contract was the delivery of technical assistance services by the Coffee Growers National Federation to the Municipality of Páramo (represented at regional level by the coffee growers' committee). Since 1959 the Coffee Growers National Federation was present in the coffee growers' municipalities delivering extension services. Afterwards, with the creation of the UMATAs, a duality was presented in the service delivery, because there were two agencies reaching out to the same producers. For this reason the provincial (departmental) coffee growers' committee of Santander proposed to the municipality of Páramo the contracting of extension services through the UMATA.

In the contract the parties involved were the municipal administration represented by the mayor and the Coffee Growers National Federation represented by the executive director of the coffee growers' provincial committee. In the contract the participation of the producers was not included.

The co-financing of the contract was initially shared by the municipality, the coffee growers' committee and the DRI Fund. This last entity eventually diminished its contribution until it disappeared, being replaced by the other two entities.

Technical assistance services were offered for several crops, among them coffee. Services were verified monthly by the coffee growers' committee and quarterly by the Sistema Nacional de Transferencia de Tecnología Agropecuaria (SINTAP), by means of monitoring records on the execution of the Annual Operative Plan (POA). The POA kept track of the number of visits, demonstration plots, technical chats, and courses offered to the producers, and supervised the programme's financial execution.

## Impact

After the introduction of the new model of contracting for extension services, the following quantifiable impacts were observed.

**1.** At an organizational level the new UMATA was able to define clear objectives within the regional policy of the coffee growers' committee, and achieved an improvement in the planning and programming of its activities.
**2.** Due to a rigorous personnel selection system on the part of the committee, it was able to employ human resources with technical capacity and leadership.
**3.** The retention of extension officers was also constant and independent of administrative changes, which guaranteed continuity in the UMATA programmes.
**4.** The semi-private feature of the model allowed a certain amount of independence to the extension agency, giving autonomy in management, and allowing more efficient use of resources (e.g. on-time payments for the execution of activities).
**5.** The new model maintained a monitoring and evaluation system that proved to be more accurate, providing a realistic view of programme activities.
**6.** As for service delivery, the UMATA-coffee growers' committee model was able to work with different products apart from coffee, enhancing technical and economic yields for these farming systems.
**7.** The model covered nearly 95% of the municipality and demonstrated a good impact as a result of the improved agricultural practices of producers.
**8.** With respect to the financial aspect, the most outstanding result was the on-time execution of payments for staff, the improved operation of the extension unit, and the progressive increase of the economic resources during its years of operation.

Apart from the easily verifiable factors of success, it was also possible to identify other impacts of more difficult quantification:

**1.** The UMATA committee model was consistent with the objectives of providing technologies with a smaller use of inputs. This was reflected in the producers' opinions of the programme.
**2.** Due to the working methods and the experience of the committee, the model was able to create a good image and institutionalized UMATA among the producers. In addition, the friendship groups methodology improved the communication and the organization of producers. The friendship groups are local organizations of producers promoted by the coffee growers' committees in order to create group cohesion, and at the same time they serve as managers of small revolving funds.
**3.** In general, the administrative aspects – control, monitoring and financing – worked better since they were within the management scope of the coffee growers' committee. This was reflected in the delivery of better services that guaranteed a better productive standard to their clients. However, community participation, due to lack of participatory methodologies, did not improve in the new extension system. Moreover, the benefits for the producers were not differentiated for men and women and there were no specific gender strategies.

**4.** For the local and national government the benefits were better delivery of technical assistance services without greater investment and with no need for a direct operative and administrative public management. For the committee the benefit was apparent in the fact that they used a working channel, the UMATAs, that had already been created in the coffee growers' municipalities. They became used to achieving greater coverage and better quality in the extension services with an efficient investment of resources.

The six key factors of success in this contracting model were:

**1.** The permanent contribution of resources by the private sector (i.e. the coffee growers' committee).
**2.** The positive image of the coffee growers' committee as seen by the producers, based on the effective and efficient delivery of extension services.
**3.** The extension methodology, which was similar to the T&V (Training and Visit) model applied previously over several years in an intensive coffee production system, with a defined and validated technological package.
**4.** The efficient administrative operation, especially on-time disbursement of financial resources that encouraged planned projects and a permanent working team.
**5.** A rigorous system of personnel selection that provided a well-paid staff of high technical capacity and leadership abilities.
**6.** The continuous and tight monitoring and control system that insured a steady execution of the goals and programmed activities.

## Sustainability and Replicability

The changes observed in the new contracting system were stable until the agreement with the coffee growers' committee came to an end. Since then the municipal administration has again taken charge of the delivery of extension services. This implies a different way of working – under a typical system of local public extension with financial shortages, bureaucratic habits and lack of continuity in the personnel.

The two implicit weaknesses of the previous model were the lack of participatory methodologies, and as a consequence of this, the absence of a participatory monitoring and evaluation system that allowed for a progressive improvement of performance. Also, the model did not consider self-financing possibilities or recovery costs schemes through a stratified charging structure based on the income capacity of the producers.

In contrast, most of the elements of the model – such as the administrative and operative management, the personnel selection system, and the control and follow-up of the service – could be replicated through training provided to the implementing institutions.

To replicate this contracting model it would be necessary to rely on a private or semi-private institution with a good image, experience, leadership, and a participatory methodology. Replication implies that financing can be framed within a national public scheme or articulated regionally through regional financing funds to which the UMATAs can apply. Participation of the private sector through farmers' organizations, non-governmental organizations or private associations of professionals in the delivery of the extension service will always be desirable.

The two key limitations to replicating the model are: (i) the financial sustainability of the model; and (ii) the absence in many regions of the country of suitable private actors for extension delivery. These structural deficiencies imply that scaling up the model to a regional or national level would require not only financial resources but the strengthening, or in some cases the creation, of private institutions in the regions to respond to the challenges outlined by the model.

## Lessons Learned

**1.** The model suggests that the contribution of financial resources from the private sector, as well as the delivery of the extension service through a non-public entity, is desirable to insure system sustainability.

**2.** For good performance in the operation of the extension agency it is necessary to have an administration and management that facilitates efficient use of resources.

**3.** Good performance of the agency demands a tight personnel selection scheme that provides a staff with technical capacity and leadership abilities.

**4.** Good results in various farming systems can be obtained with clear and defined technological packages.

**5.** To achieve financial sustainability requires a combination of self-financing strategies (i.e. the sale of inputs or parallel services), costs-recovery schemes (such as farmers stratified payment) and the decentralization or regionalization of the national system of financing.

**6.** The momentum of the model was achieved when a national extension programme – focused on the local level with available resources from the national budget – was merged with the experience of a farmers' organization that also possessed commercial experience and was able to make contributions to finances.

Finally, for those interested in this institutional experience, further information can be acquired from the coffee growers' departmental committee of Santander, with headquarters in the cities of Bucaramanga and San Gil. The committee also provides training in the main methodological elements of the model. For the moment, experience indicates that the financial sustainability of the model remains a problem to be solved.

# Chapter 12

# Trinidad & Tobago: Contracting for Extension Communications Services – the Hibiscus Mealy Bug Information Campaign

Joseph Seepersad and Wayne Ganpat

## Context

This chapter discusses the experiences of the Ministry of Agriculture in Trinidad in using commercial mass media agencies to produce and disseminate extension messages on the identification and control of the Pink Hibiscus Mealy Bug (HMB). The advent of this insect pest in Trinidad in 1995 forced the authorities to deal with the use of mass media in a way that they never had to before. The HMB had already wreaked havoc among crop and non-crop species in neighboring Grenada so its destructive potential was well known.

Early efforts at eradicating the pest in Trinidad when it appeared in isolated pockets were not successful and it was soon evident that the problem had to be tackled as a matter of urgency on a national level. Since the pest had a wide range of host species, the drive was aimed at the public. A massive public education programme had to be undertaken initially to enable the public to properly identify the pest and to use short-term control measures. Later efforts were aimed at educating the general public, including farmers, on biological control, the identification of the pest's natural enemies and the need to preserve them.

The Extension Division has a media unit responsible for producing radio and video programmes and audio-visual material to support field extension activities. The unit is staffed with one professional along with support staff. Due to funding restrictions, however, the equipment could not be replaced or updated as often as needed and was thus somewhat unreliable. While a more modern central facility was available to service all government departments, the demands for its services far outstrip its capacity.

In this situation of crisis it became clear that in-house facilities did not have the capability to produce the amount of material needed in a short space of time. Getting material of good broadcast quality was also an important consideration, since the wide availability of cable TV etc. has led to a rather sophisticated public mass media audience. Evidently, public service announcements and other types of materials had to measure up to certain standards to catch the public's attention. The Ministry of Agriculture therefore decided to enter into contracts with commercial media houses to produce video materials for broadcast TV and a jingle for radio.

Decisions regarding the role of the media houses vis-à-vis staff in the Ministry of Agriculture were made carefully. The real bottleneck was the technical aspects/quality of the production of the videos and therefore the media houses were "hired" for that purpose. The technical content, scripts shot lists etc. (the pre-production aspects) were developed by the Ministry's staff. The media houses were contracted for the production and post-production operations – filming, editing etc. Similarly, most of the materials for radio programmes were developed in-house. The production of a suitable jingle, however, was contracted to a media house since this required certain skills and specialized resources. Print materials were developed and produced within the Ministry, while posters were sent to commercial print shops for mass production. These, too, required appropriate facilities since they were designed in colour and needed to show clearly the symptoms of attack, the appearance of the insect and its predators etc.

The media houses were selected through a tendering process following the normal rules for government contracts. Interested companies submitted bids from which the final selections were made. The process was managed and executed by the Extension Division of the Ministry of Agriculture. Since the requirements for the products were pretty specific, no real problems were encountered in getting the desired output.

The Ministry also broke with normal practice to ensure that the information reached a large percentage of the population as quickly as possible. Normally government programmes were aired in the free slots provided by the broadcast media. The available slots, however, were not set at peak viewing or listening times and so the government decided to pay to ensure that information was delivered at "prime time" as frequently as needed. The stations and times selected were based on reports of various surveys.

## Impact

There was a high level of awareness in the national community about the HMB. It was popularized in local songs, was the subject of debates in parliament, the theme of editorials and several letters to the editors of newspapers. That the overall programme dealing with the HMB was a success has been acknowledged widely and several countries in the region have looked to the Trinidad experience to guide their efforts. The success was no doubt due in part to the timely and effective use of the mass media to focus national attention on the seriousness of the threat and to mobilize efforts to combat the problem. One study conducted in the southern region of the country (Dowlath and Seepersad, 1997) confirmed the importance of television as an information channel for all aspects of information about HMB – over 50% of the respondents in five out of the six aspects examined indicated that television was the most important source of information.

In examining factors contributing to the success of the contracting arrangements, it is difficult to isolate the individual contributions of various factors resulting in overall success. Many things came together well. The urgency and seriousness of the problem prompted the government to take quick action and to be favorably disposed to using any approach that would get the job done. The technology that was eventually adopted, i.e. biological control, was effective and appropriate judged from several perspectives. Initially the message was "Spray, Cut and Burn" but those recommendations became very difficult to implement as the problem escalated.

The approach to the information/education campaign also seemed to have been well planned and executed. In addition to the mass media thrust, a variety of print media were used – posters, flyers, and fact sheets. Extension staff used group meetings with householders, schools etc. and individual visits in "high risk areas" which had large populations of the pest. Exhibits were set up in markets, malls and other areas frequented by the public.

The contracting arrangements allowed the mass media to play a critical role and enabled quick and timely intervention. The decision to pay for prime time was a key factor also, since it would not have done any good to have well prepared materials without getting to the public. The fact that the extension staff were able to do a lot of script preparation cut down on money and time. The media houses do not have in-house personnel nor the experience to deal with the subject areas. The scripts etc. also provided a clear framework within which they could operate so that there were no real problems regarding the outputs matching the expectations of the contractor. Nevertheless, the staff in extension kept in constant contact with the producers to ensure time deadlines and quality requirements were on track. Payments could therefore be made without long delays that could occur if the output was not satisfactory.


## Sustainability and Replicability

The experience has paved the way for similar types of arrangements in future, particularly since the overall programme, was a widely acknowledged success. If the efforts were not that successful then it would be more difficult to convince government to go that expensive route again unless unsuccessful and successful components were identified clearly. The urgency and seriousness of the problem no doubt "forced" the government's hand in this direction and so it is likely to be regarded as a valid model mainly for crisis situations. Nevertheless, the experience can be used to canvas for similar arrangements, though not of the same intensity, during the regular programming cycle. A good approach might be to have government make provisions for this eventuality in the regular budget so that one does not have to make a special case later.

In considering the transferability of the "model" obviously there must exist commercial media houses capable of doing the job. As demonstrated in this case, it does not matter too much if they do not have the technical expertise once that is accessible somewhere in the knowledge system. It will help tremendously as well if extension itself has the capability to "package" the technical material for the mass media. Perhaps most important of all is that the "leadership" in the system should have a good understanding of the overall picture – the role of the various channels, various "models" of linking the private and public sector etc. In situations of crisis such expertise can be sourced from international agencies if it does not exist locally, but it is best to develop in-house talents for continuity of effort. Of course, "nitty-gritty" things must not be overlooked, e.g. impartial tendering procedures, quick processing of bids and the ability to make payments on time once the job is completed etc.

## Lessons Learned

As already mentioned, this case shows the way forward for using the mass media effectively in extension programmes even though the organization lacked proper in-house facilities. In situations like that discussed, it may even be the most cost-effective approach, since it will be expensive to maintain a media unit in the long-term to meet an occasional or seasonal heavy demand. This does not mean that an in-house media unit is not necessary; there is no doubt that the staff's experience and insights in packaging materials for the mass media played an important role.

Extension methods in the past have been dominated by individual approaches such as the farm visit, and the call for increased farmer contact is becoming more strident. Mass media possesses the potential for increasing extension's reach, but its consistent and effective use has proved to be too much of a quantum leap for most extension organizations in developing countries. In the Caribbean, a fair amount of effort has gone into establishing communication units to produce teaching and media materials in-house, but reduced funding from public and external sources has slowed their momentum. The challenge has become greater with the greater accessibility of the public to cable television so that high quality materials must be produced to catch people's attention. Clearly, the idea of "Contracting for Extension Communications Services" deserves serious consideration as an approach to harnessing the mass media for extension, in spite of the existing constraints.

# Reference

Dowlath, P. and Seepersad, J. (1997) The role of information channels in communications: the case of the Hibiscus Mealybug Campaign. Paper presented at the Ninth Seminar on Agricultural Research and Development, National Institute for Higher Education, Research, Science and Technology (NIHERST), Couva, Trinidad, 26-28 October, 9 pp.

Chapter 13

# Vietnam: Contracting for Extension Training in Participatory Planning Methods

Dinh Duc Thuan, Nguyen Ba Ngai and Bardolf Paul

## Introduction

From January to December 1997, a training contract was executed between the World Food Programme (WFP) Project 5322 "Smallholder Forestry Development in Five Northeastern Provinces of Vietnam" and the Social Forestry Training Centre (SFTC) at the Vietnam Forestry University (VFU). Helvetas Vietnam, a Swiss non-governmental organization (NGO), provided technical assistance in planning, supervising implementation, and documentation. The Ministry of Agriculture and Rural Development was the implementing agency for WFP 5322, and therefore approved and coordinated the implementation of the contract.

Deforestation is a major concern of the Government of Vietnam, because of its negative impact on livelihoods and the environment, and WFP 5322 aimed to assist 51,000 poor households to develop 51,000 ha of allocated forest land (1 ha per household). These households were chosen from 940 villages located in 157 communes in 22 districts in five northeastern provinces. An average village has approximately 54 participating households.

Under the project, each village and every household is expected to prepare a plan for development of their allocated forest lands, with project assistance. Over a period of 4 years, extension staff and village extension farmers were expected to assist villagers to prepare 940 village plans and at least 51,000 household plans. A participatory planning process was adopted for this purpose, and also was the means by which project participants were selected. Each village established a Village Management Committee, chose one demonstration farmer, one extension farmer per 25 participating households, and one village tree nursery worker for each 10,000 tree seedlings to be produced.

## The Contract

In order to achieve the above targets, training had to be provided to extension staff and village extension farmers. This was the purpose of the contract between the parties – to train trainers in a simple method for participatory household and village planning. The planning system prior to the contract was based on a traditional, top-down, centralized approach. The parties involved recognized the need to change the approach to incorporate a more decentralized planning mechanism. An important part of the new approach was the central role of a village-based extension network that would bridge the gap between the weak and understaffed government system and the needs of the farmers. In summary, the three main obstacles that had to be overcome were:

1.  Lack of an understanding of participatory planning approaches and methods.
2.  Lack of planning staff who could function as extensionists.
3.  Lack of an existing grassroots extension system.

The contract formulation was rather unique because the original concept from the WFP was to hire an external consultant for three months. Helvetas was approached to provide the consultant because of their expertise in this area, and instead suggested getting the SFTC at the VFU to provide the bulk of the training services, with assistance and supervision from one of Helvetas' foreign experts. A contract was formulated between WFP, MARD, Helvetas and SFTC, in which the parties agreed to the above arrangement. The formal contract was signed between UNDP/FAO and Helvetas.

The form of the contract was a "Reimbursable Loan Agreement" in which Helvetas agreed to loan one of its long-term advisers to FAO for three months in exchange for a fee. The bulk of this fee (approximately 90%) was used to sub-contract the services of the SFTC to implement the contract. The remainder was used to cover time spent by the Helvetas adviser. Payments were scheduled in two tranches – one in the middle and the other at the end of the contract, based on completion of training activities and acceptance of a final report.

## Impact

Obviously, this type of indirect contractual arrangement is not ideal, but given the circumstances at the time it was a very satisfactory and expeditious arrangement, for the following reasons:

1. For SFTC: the Training Centre had been working with Helvetas since the late 1995 providing similar contractual services to other bilateral, multilateral and NGO projects. One of the main objectives of contract work is for staff to get hands-on field experience that they then could use in their curriculum

development and teaching work. This contract gave SFTC additional experience in collaborative and cooperative capacity building in participatory planning and extension training, as well as much-needed revenue to supplement meagre salaries and skimpy operational budgets.

**2.** For WFP: the involvement of Helvetas provided a level of security that the provision of services would be professional and satisfactory. WFP had no prior exposure to the SFTC, and were not willing or able to contract directly with them.

**3.** For MARD: the Ministry achieved much higher value from the investment, because under this innovative arrangement they received a much greater number of consulting days, which enabled them to cover a much greater area in a shorter period of time.

**4.** For Helvetas: the contract helped further Helvetas' capacity building objective for the SFTC, providing a real life situation for the teachers to practise in and gain experience from. Moreover, there was no extra cost to the SDC project.

**5.** For Trainees: trainees at all levels – province, district, commune and village – took part in a unique "process training", in which training was combined with on-the-spot coaching and guidance during the implementation process. This enabled this diverse group to gain new skills and expertise in applying participatory planning and extension methods for forestry and agroforestry development.

Contracts are not a new device in Vietnam; however the overall tendency is for agencies to use internal resources and to avoid paying for external services. In part this is a reflection of a history of scarce resources, so that outsourcing expertise simply was not possible. Because of this, institutions tended to function in comparative isolation. In this situation, the contract was funded externally, so the use of external expertise was not an issue. What was unusual in this case was getting agreement to use the funds mostly for national instead of expatriate expertise. So one of the impacts was exposure to the possibility for more creative and flexible use of contracts.

One aspect that was not appreciated by the Ministry was not having direct control over the Vietnamese training team, who were sub-contracted with Helvetas. This meant that the Ministry officials did not have the authority to give instructions to the team, which from the training team's perspective was positive, but not so in the eyes of some Ministry staff.

The contract had both quantitative and qualitative targets. The quantitative ones were very easy to measure, because very clear records were kept of all activities – what happened, when, where, and who was involved. The qualitative aspect was experienced by all parties concerned, through direct participation in field activities and in follow up monitoring and evaluation. Periodic reports were made to review and assess training and field activities, and their impacts. The contract required an end-of-contract report and self-assessment.

As with many project-based initiatives, the contract was not designed to achieve institutional change or capacity building outside of the objectives of the project. This is short sighted, because it means that the impact is short term and bound by the tenure and terms of the project. Exposure to new approaches and ways of working can be very useful in opening up people's minds, especially when it has a practical application that can be seen and experienced directly. Certainly there was a lot of that in this case; however, unless institutions are prepared to apply this experience in their other areas of work, it makes for little long-term impact, which is what happened here.

However, overall the following key impacts were achieved as a result of the process:

**1.** Development of an innovative and effective mechanism to spread a participatory planning and training approach quickly over a relatively large area.
**2.** A more sustainable result through the direct involvement of all parties, especially the farmers, in the training and planning process.
**3.** A relatively low cost mechanism for training, planning and implementation.
**4.** The direct linking of training to field application.

## Sustainability

How sustainable are the changes that were brought about through this contract? As mentioned above, when an intervention like this is bound within a project, it tends to ignore to some extent the wider questions and issues outside of the terms of the project. In the case of WFP 5322, there has not yet been a formal evaluation, so it is too early to assess the permanency of change that may have taken place.

On the ground, plans have been made at village and household levels, nurseries are established, and trees have been planted. That is all very good and positive. And in the case of the trees, has a good prospect for sustainability as well. Some of the new approaches involving participation place added budget demands on Vietnamese government institutions, so there is some concern about their sustainability in the long run. In fact, once the initial training was finished there was a delay in implementation, until the government approved field allowances for newly trained staff. This was a factor that was not anticipated in the project planning.

The chances for sustainability, however, are high in terms of the mechanisms that were introduced because they provided a much more realistic and effective means for villagers and extension staff to interact and be mutually supportive. Planning and implementation were based on actual needs and local contexts, and are much more likely to be sustainable because of the resulting effectiveness.

# Replicability

In terms of replicability, the same arrangement and process proved itself to be very workable. The initiative was on a relatively large scale. In fact it was the first training exercise in extension that had been carried out on such a scale and within such a short time frame – 75 training events in six months, involving over 500 direct trainees and several thousand farmers as indirect trainees. It would have been more comfortable to have had more time, but there was no significant compromise on the training itself due to shortage of time or intensity of the time schedule. This may not have been the case with the government staff, because they were engaged in an exercise outside of their normal routine, and there were very likely impacts on their routine work activities.

One of the major constraints on upscaling is the limited numbers of qualified trainers. There are not too many people around and available with the requisite skills and experience to provide this type of training. It requires fairly specialized sensitivities and is quite demanding on the trainer's resources. This would be the first and most significant constraint on replicability. Secondly, it would be much better for replicability to change from a project to an institutional basis in which both leadership and mainline staff are committed to supporting a process of change. This implies an appropriate commitment of resources – material as well as human – to support widespread replication.

# Lessons Learned

As already mentioned, the contracting process is not new in Vietnam, especially for "hardware" applications such as construction or machinery. There has not been a lot of contracting for "software", i.e. services and personnel, especially in extension. The extension system is fairly new; it came into being formally in 1993 and is still somewhat rudimentary in many places, especially in more remote mountainous areas, such as where much of this training contract was implemented. So in general, there is a lot of learning to do in building capacity and expertise.

Training capacity does not exist within the extension system itself, and most staff are completely untrained and inexperienced. Therefore there is an enormous need for external assistance in helping to build capacity and capability. Where should this come from? This fundamental question has not been adequately answered yet. There are a lot of ad hoc initiatives, with many foreign funded projects doing what they can to fill the gap. But these are basically short-term solutions, lacking sufficient forward thinking, follow up and overall coordination.

The contract work that the SFTC has been doing provides a useful approach that is worth studying. The combination of a long-term expatriate adviser coupled with the SFTC training team has proven to be a good way to work within the Vietnamese context, maintaining quality while delivering quantity. At all levels everyone needs to learn, and the best way is through action, working together and sharing experiences. Thus the combination of the SFTC working directly with local extension institutions, and with community leaders and villagers, is very effective in the spread of learning and capacity building. Cooperation also establishes good feedback loops between the farmers, extension workers, extension organizations, the training organization, and in this case, the Ministry. Therefore, it can be very beneficial to involve a multiplicity of institutions – international, national, local, governmental and non-governmental – with each taking an appropriate role and responsibility, in a collaborative constellation with local people.

The following tools and mechanisms were mainly used to initiate and guide change:

1.  Using a "process-based" training of trainers approach.
2.  Combining training with "learning by doing".
3.  Open and collaborative negotiation between three main parties – foreign experts who introduced methodologies, Vietnamese specialists consolidating and grounding methodologies, and extension staff working together with local people to provide appropriate levels of support for planning and implementation.

Maintaining the momentum is a major challenge that faces two obstacles in the current situation: insufficient budget to support this type of interaction, and the inclination toward non-cooperation and institutional isolation. Without resolving chronic budget constraints, it is very difficult to initiate participatory approaches; and these approaches are the most effective mechanisms for breaking down barriers between institutions.

## Lessons for International Organizations

For international organizations to learn from this type of experience, it is essential that their staff examine real life, practical proposals and plans, and develop expertise about how to provide advice, and more importantly, how to take people through similar learning and development processes. There is no substitute for first-hand experience, and every situation is different. Basic principles can be applied, but how they are applied is the critical factor, and it always requires the hand of experience to achieve success. In Vietnam at this time, what appears to be best is when Vietnamese experience can be combined with well-seasoned expatriate expertise.

# Reference

Paul, B. (Helvetas) (1997) *Mission Report: Training and Extension Consultancy for UNDP/FAO Project 96/014 Building Capacity for Smallholder Forestry Development.* Hanoi, Vietnam.

Section Five

# FARMERS CONTRACTING FOR COMMERCIAL ADVISORY SERVICES

# Chapter 14

# Portugal: Contracting for IPM Extension by the Association for Viticultural Development in the Douro Valley

Artur Cristóvão, Fernando Alves and Timothy Koehnen

## Context

The Douro Valley is a world renowned area of wine production, due particularly to the character of its fortified wine, sold with the label "Port Wine" and produced since the 18th century. In recent years, the quality of its table wines has also gained recognition, and production has been steadily expanding.

The Douro Valley is an area mainly of small and very small farms, the average area being only about two acres (80,000 acres, 38,000 producers). Medium and particularly large farms form only a minority, yet have considerable importance and power, due to their involvement in the wine industry and marketing. Most small farmers sell their grapes to cooperatives and private wine-making estates located in the region.

The Association for Viticultural Development in the Douro Valley (ADVID) was created in 1982 by a small number of owners of large wine estates. ADVID's mission is to promote technological innovations, in order to improve the quality of grapes and wine. To accomplish this, ADVID offers farmer-training courses, develops on-farm trials, organizes field demonstrations and provides technical support through individual visits and information bulletins.

ADVID has been involved in plant protection experimentation and extension ever since its foundation. However, Integrated Pest Management (IPM) work started only quite recently since the knowledge base was quite restricted, policy incentives did not exist, and human and other resources were very scarce. Since the early 1990s, a group of 12 mostly large farms, totaling about 2,200 acres, has received extension help to introduce IPM practices.

With the aim of promoting sustainable agricultural practices, the European Union, under the auspices of the reformed Common Agricultural Policy, launched the so-called Agro-Environmental Measures. These measures were implemented in Portugal after 1995, and a strong IPM component was adopted. The number of farmers involved in IPM programmes has been growing quite rapidly since 1994. Between 1994 and 1996, 91 training courses were offered, which were attended by 383 technical staff and 1,462 farmers. The number of participating farmers surpassed clearly all official expectations, and has since grown further from 5,254 in 1998 to 8,509 in 1999. The investment in this field, in the 5-year period between 1994 and 1998, was four times the amount initially planned. In June 1999, 41 associations were working on IPM extension and new associations were also actively planning to get involved, namely by recruiting and training extension agents, in order to obtain the required official certification. A total of about 200 technical staff was already working in the whole country (Amaro, 1999: 63-66).

IPM measures were designed to be implemented with strong support from local associations. In this way, many existing associations became involved and many new ones were also created with this aim in mind. Essentially, they saw an opportunity to obtain resources and start or expand extension and other field activities. That was the case of ADVID, which saw the IPM programme as a unique opportunity to acquire new logistical means, hire staff, enlarge the scale of intervention, involve more members in IPM and widen the scope of its work to all dimensions of vineyard management.

The IPM contract is mainly a technical service contract and involves three partners: a farmer (with a minimum of 2 acres), an association, and the Regional Agriculture Services. Under the contract: the farmer receives a subsidy per acre, funded by the European Union (75%) and the Portuguese Government (25%). Farmers have to follow the recommended IPM practices for at least 5 years (the subsidy decreases as the acreage increases) and they have to attend a 35 hour IPM training course and, if deemed desirable, other IPM continuing education activities. Farmers have to be a member of a certified IPM Association, such as ADVID, to which they pay 25% of the subsidy received. The association then provides training and extension support to the farmer (IPM courses, farm visits, demonstrations, technical meetings, etc.), and is also required to control the effective application of the IPM rules. The Regional Agriculture Services, part of the Ministry of Agriculture, Rural Development and Fisheries, supervises the programme, visits the farmer and checks the vineyard management practices. If IPM recommended practices are not being followed, both farmer and association are penalized. For example, the association may lose its certified status as an IPM extension provider.

Farmers incur eight obligations under the IPM contract (ADVID, 1997):

1. The farmer must be a member of a recognized association.
2. He must attend an IPM training course in the first year of the project.
3. He must fulfil the approved IPM rules.
4. He must use exclusively the accepted IPM phytopharmaceutical products.
5. He must adhere to the rules of the contract signed with the association for a minimum of 5 years.

6. He must follow, with the extension agent, the biological cycle of the crop's main enemies, undertaking periodical risk assessments and evaluations of damage, and registering the information in a field log.

7. He must provide the control agent with complete and correct information on IPM plots, and to facilitate access to soil, plant or fruit samples for analysis of residues.

8. He must commit him/herself, by signing the contract, to practice the principles and rules of IPM in the registered plots.

The IPM associations also commit themselves to eight obligations (ADVID, 1997):

1. To maintain IPM activity for a minimum of 5 years.

2. To hire a number of qualified technical staff, according to the size of the registered IPM area.

3. To provide technical support to the associates, according to the contract.

4. To control the activity of the associates, once a year and 2 weeks before harvest.

5. To provide official institutions with a list of IPM farmers.

6. To offer initial and continuing IPM training courses.

7. To make available and promote use of an appropriate field log.

8. To promote commercialization of IPM products.

## Impact of IPM Extension Contracts

Since 1997, four new technical staff have been hired by ADVID, to act as IPM extension workers. The number of participating farmers grew from 12 to 130 and the total area from 2,200 to 6,000 acres. New types of ADVID members are involved in IPM. In the past, there were only large estates with an average area of 183 acres; today there are also wine cooperatives and others with smaller holdings (averaging 32 acres), the smallest having only 10 acres. Small farmers are still a minority.

Impact studies, involving external evaluation, to analyze aspects such as productivity or quality of grapes and/or wine, have not yet been done. Nevertheless, it is clear that the IPM contracts have clear-cut results in terms of farmers' practices, health risks and limiting environmental damage. These improvements include a more careful choice of pesticides and other chemicals, better application practices, fewer cases of poisoning and other health problems, and possibly decreased contamination of soils, water and air. Other impacts are not as successful and deserve further attention. For instance, environmental consciousness is still quite limited since the economic incentive is the major change factor; methodologies to evaluate the risk of pests and diseases at the farm level are not yet mastered by most participants; and the resulting wines are not yet certified as IPM products.

As for the impact on ADVID, it is recognized that this association has gained considerable technological knowledge and know-how on IPM matters. It has gained field experience. Staff have participated in advanced training. And it has strengthened relationships with a larger number of associated members and other involved actors, both in the region and abroad.

ADVID has derived major benefits from the IPM programme. Greater and better knowledge of IPM principles and techniques has been gained, along with more knowledge of the choice of pesticides and other chemicals, and related application techniques. It has improved the organization of vineyard operations, with positive impacts on farm management and production costs. It has clearer perspectives of improved product quality, which has a potentially positive impact on sales, due to consumers' increased health awareness.

The success of the programme, namely in terms of the growing number of farmers involved, seems to be linked mainly to the economic incentives, that is, the subsidy granted per unit of registered acreage. However, other factors deserve to be mentioned. These include: (i) ADVID's previous experience in IPM and the intensive extension, training and experimentation work developed since 1997; (ii) the increased trust established between extension workers and participating farmers; (iii) the rigorous definition and application of a set of IPM techniques well adapted to regional agro-ecological conditions; and (iv) the provision of quality certification of the wines produced with IPM grapes.

## Sustainability and Replicability

Since the programme is of recent origin, it cannot be considered sustainable at this stage. In order to accomplish sustainability, farmers need to have more technical capacity to make decisions on IPM matters (knowledge related changes); their environmental consciousness needs to be strengthened (culture related changes); and the economic effects of the practices adopted have to be demonstrated clearly, namely in terms of reduced production costs and increased incomes (economic changes). On the other hand, contracted IPM extension needs to be coupled with supportive legislation and regulations on pesticides' sale and use, environmental education initiatives, public extension education and research efforts, and appropriate agricultural education, both at vocational and higher education levels.

Similar types of programmes have been established in Germany, Italy, Spain and Switzerland, along the lines of a similar agro-environmental policy and funding approach. In Portugal, they have also involved fruit producers (plums, peaches, apples, pears and citrus), and there are plans to include olive producers as well. Basic ingredients for replicability and success, as already underlined, are the financial package and economic incentives associated with IPM, and the existence of sufficient organizational and technical capacity on the part of the associations. During a transition period from a more traditional extension system to a contractual arrangement, the associations need to be adequately supported, namely with training opportunities for the new technical staff and parallel information efforts.

The IPM programme can be scaled up (more farmers, more acres), but only within distinct limits. Careful attention needs to be paid to the ratio of farmers, or acres, to extension agent, as a more personalized relationship is essential, so that regular and close technical follow-up can be done and a good learning environment created. In ADVID's case, each agent works today with about 22 farmers, and this number is not expected to rise to more than 30. On the other hand, a larger operation will make it more difficult to guarantee the required quality standards.

In other regions, experience shows that most associations also work with a single crop and that the area per extension agent varies quite a lot, in the case of vineyards, 1,000 acres being the maximum (Amaro, 1999: 67), a value mirroring ADVID's situation.

Experience also shows that scaling up can be better accomplished through the multiplication of IPM associations in the same region. In the area studied, five such associations exist today, all working with vineyards, some of them reaching more farmers and vineyard acreage than others. It is desirable that they coordinate activities, for instance in the field of training, but this has not been happening so far.


## Lessons Learned

From ADVID's experience, a number of lessons can be drawn, most likely in line with what has happened with other associations in the country.

**1.** There is a strong interest in IPM on the part of farmers, due mainly to economic considerations and not so much because of environmental concerns.
**2.** Serious IPM work requires strengthening institutional contacts, both within the country and abroad, in order to promote exchanges of knowledge and experience (with universities, research centers, associations of the same or similar type, related business enterprises, for instance dealing with equipment and pesticides, etc.). Associations need to develop their technical and organizational capacity to perform responsibly and with high quality.
**3.** Contracted IPM extension requires improved linkages with the Regional Agricultural Services and other regional/local institutions, through mechanisms such as joint planning meetings, leading to coordinated strategies and plans.
**4.** IPM extension has to promote closer relationships with neighboring farmers, for example through informal groups, to stimulate an open discussion of IPM farm practices and comparisons with the farmers' traditional intuition and local knowledge. This will potentially help to increase the present farmer/extensionist ratio and decrease inequality.

Farmers' training courses and extension visits tend to promote knowledge exchanges, questioning the common practices and gaining more of the critical capacity needed to assess the traditional ways of doing things and, in this sense, should be carefully designed and implemented.

Contracted IPM extension has established a foothold within the Douro Valley area. However, holdings with less than 2 acres may require an alternative strategy and/or type of incentives for IPM. The instruments or mechanisms to initiate and guide changes in this field, in this specific context, were multiple; the following are particularly important:

**1.** Economic incentives, that is, the subsidies per registered acre.
**2.** Institutional support, that is, the mediation of the process and support through local associations, paid by farmers to deliver specific extension and training activities.
**3.** Technical procedures, that is, a clear and appropriate set of IPM rules and practices, basis for farmers' decision-making and action.
**4.** State control, that is, monitoring of farmers' IPM performance, undertaken by the Regional Agricultural Services.

State control has been considered quite inefficient according to national IPM experts, who claim that technical monitoring of IPM is practically ignored, with official agents being more concerned about administrative aspects such as confirming acreages, filling in of field logs, and the nature and training of technical staff (Amaro, 1999: 68). Changes in this regard are critical in order to maintain momentum in this process, particularly considering the growing number of farmers and associations involved. At the same time, legislation and regulations on pesticide sales and use need to be defined, improved and applied.

The future development of contracted IPM extension may include: (i) progressive subvention schemes, starting with contracts for a basic IPM programme and evolving according to the number and complexity of environmental measures adopted by farmers; and (ii) selective support to associations, according to evaluation of the technical and educational capacity to effectively implement the programme, and not according to size of membership.

The contact with local associations and regional supervision institutions seems to be the first step to take in order to obtain more information and to start any training of international staff or other technical agents. ADVID has expressed its willingness and readiness to organize field visits and training opportunities.


## References

ADVID (1997) *Protecção Integrada da Vinha*. Boletim Informativo no. 10.
Amaro, P. (1999) *Para a Optimização da Protecção Integrada e da Produção Integrada até 2006*. Universidade Técnica de Lisboa, Instituto Superior de Agronomia, Lisboa.

# Contacts

Eng. Fernando Alves, Director of ADVID and responsible for the IPM Programme, Rua José Vasques Osório, 62, 5050 Peso da Régua, Portugal
Tel: 351 254 312940
Fax: 351 254 321350
Email: advid@mail.telepac.pt

Dr Isabel Escudeiro, Director of Rural Development Services, Direcção Regional de Agricultura de Trás-os-Montes, Rua da República 197, 5370 Mirandela, Portugal
Tel: 351 278 257147
Fax: 351 278 257603
Email: isabel.escudeiro@dratm.min-agricultura.pt

# Chapter 15

## USA-Louisiana: Private Row Crop Consulting and Implications for Interdependency among Private Consultants, Extension and Farmers

John Barnett and Satish Verma

### Authors' Note

The information in this case study is based on an analysis of in-depth interviews with five private consultants working in major cotton production areas of Louisiana. These consultants have experience in the consulting business ranging from 15 to 50 years. In addition, three cotton farmers located in three geographic locations in the state responded to a mailed survey. They have large operations – 4,000-9,000 acres – and employ full-time consultants for a broad range of services. The authors have relied on their personal knowledge and experience of the cotton industry and the extension system in Louisiana to interpret the data collected and to add their own perspectives.

### Context

Private crop consulting services in the cotton industry of Louisiana, USA, began in 1949 in Caddo Parish (same as county in other states) in the northwestern part of the state. These services spread slowly to northeast and central Louisiana, the other major cotton growing areas of the state.

Prior to 1949, the major provider of agricultural production and management information to farmers was the Cooperative Extension Service. Chemical company representatives and aerial applicators also gave farmers information and advice on cotton insect control and treatment. Aerial applicators, in fact, sprayed and dusted cotton fields in the early 1920s. Farmers engaged aerial applicators to begin spraying or dusting their fields at the first sign of insect infestation and followed a set schedule for the remainder of the season. Neither farmers nor aerial applicators had a scientific basis for applying insecticides. In the late 1940s, professional cotton scouts began to check cotton fields for insect infestations and make science-based control recommendations. In the 1950s and 1960s, "generalist extension agents", who had the responsibility of helping farmers in a broad range of subject matter, and chemical company representatives, who had limited technical education, experienced difficulty meeting specialized needs of farmers raising crops in an increasingly complex agricultural system. Consequently, extension agents and chemical company representatives began to be less directly involved in providing specialized information and advice. This led to the emergence of a cadre of professional, specialized consultants.

The growth of the consulting business has seen consultants branching out into several areas. Today's consultants, although still a primary provider of pest management advice, offer a variety of professional services, including budgeting, farm planning, land evaluation, governmental regulation compliance, selection and placement of crops, variety selection, use of growth regulators, harvest aids, and soil fertility.

Professional consulting is a fee-based system. The farmer is charged according to the services provided. In contrast, extension advice and recommendations have always been free. Chemical company recommendations have also usually been free, although a sales pitch for company products is made.

Chemical company recommendations are not as prevalent as they once were, but extension continues to be a vital link with farmers and consultants. Extension agents and specialists work year round to provide the latest production, management, and marketing information and recommendations. However, consultants are usually viewed by producers as the first line of on-farm defence in insect control.

Currently (1999), there are over 100 professional consultants in Louisiana providing contractual services to row crop farmers, including a majority of the 3,193 cotton farmers in the state. That same year, the estimated cotton acreage was 609,885 acres, and the estimated value of the crop, including seed, was nearly US$260 million. Consultants are tested and certified in specific expertise areas by the State Department of Agriculture before they can practice.

It is estimated that 85-90% of the cotton acreage in Louisiana is under contract with a professional consultant. The contract is between the consultant and family farmers, partnerships, or corporations. Fees charged vary from US$4.50 per acre up to as much as US$16.00 per acre depending on the services provided. Usually, fees are paid up front or in two or three installments. Fees are not negotiable, but growers can add desired services for additional fees as the season progresses.

Surprisingly, many contracts are still verbal agreements between the parties involved. Agreements stipulate the responsibility of the consulting firm to: (i) conduct regular field inspections using state-certified consultants; (ii) submit pertinent field inspection reports and recommendations; and (iii) protect client confidentiality. Producers have to permit consultants access to their fields and follow recommendations. Fees for consultant services in different commodities are stipulated.

## Impact

The introduction and expansion of the private consulting business in cotton has had wide-ranging economic, environmental, and technological impact at all levels of the industry.

From a state perspective, the most significant impacts over the last four decades are: (i) the increase in cotton acreage under contract; and (ii) the increase in number of cotton farmers engaging professional consultants to perform a variety of services. In 1999, 90% of the cotton acreage, or 550,000 acres, was under contract to consultants. At an average of US$10 per acre, the gross income of the consulting business from cotton alone was over US$5 million. Employment and other economic benefits from consulting are also considerable. Again, in 1999, all producers cultivating 1,000 acres or more employed a consultant. For cotton farmers, this means more and better use of their time to focus on other critical farm operations such as marketing, programme compliance, and labour, equipment, and financial management.

For consultants, increased credibility and farmer confidence in their services are evidenced by steady return business from farmers over the years, as well as the high level of farmer adoption of consultants' recommendations (95%). Apparently, cotton farmers are satisfied that well-trained, certified professional consultants are bringing relevant, specialized technical information and recommendations using up-to-date research to help them deal with problems they face in increasingly complex and specialized agricultural systems. Although these farmers have a financial stake in the consultant, they have to receive economic returns in excess of their investment in consultants to continue to use them. Apparently, this has been happening.

Besides increased income, significant benefits to cotton farmers who use consultants and follow their recommendations are: (i) assuring that an expert with a "good pair of eyes" is looking out for them; (ii) achieving overall management efficiency; (iii) projecting a progressive, credible image to lending institutions; and (iv) viewing consultants as a management tool, like any other tool, for improving overall production and management.

Another significant impact of private cotton consulting is protection of the environment from the indiscriminate use of chemicals by farmers and aerial applicators. The positive effect on the environment of using science-based pest management and production operations, or best management practices (BMPs), in the cotton industry as recommended by trained professional consultants has been considerable. Aerial applicators, who had been spraying and dusting cotton fields as requested by farmers based on visual observation and a pre-set schedule, had now to follow recommendations of consultants who checked fields for insect counts, and the presence of weeds and diseases. Aerial applicators, who were essentially sales persons for chemical companies, eventually ceased to be a major source of advice to farmers. They are now pilots. In this change, a simplistic, environmentally degrading system of plant health insurance has been replaced by a more elaborate, scientific, and environmentally sound system. Benefits in preserving soil and water quality have been significant.

For the most part, the private cotton consulting business has strengthened extension-consultants-farmers relationships. These relationships are viewed by the parties concerned as synergistic, cooperative, and mutually beneficial. However, tensions can occur when: (i) consultants and farmers undermine the credibility of extension agents; and (ii) extension agents regard consultants as a threat to their farmer client base. On the other hand, the reasons for the strong relationship are obvious. Extension, backed by research, continues to be a source of unbiased, up-to-date, timely information and training opportunities for both consultants and farmers. Extension agents work closely with farmers and consultants to meet specific educational and information needs. Consultants rely on extension-research for current information which they can use in their consulting business with farmers. Farmers engage consultants to assist them in a variety of technical areas, depending on need, affordability, and relevance to their operations. At the same time, they continue to participate in extension education programmes and are advocates of extension, testifying to its credibility and value.

Major benefits of the private consulting business include:

1. Farmers are making better production and management decisions.
2. Farmers are kept informed by consultants about the latest technology, products and services. Thus, they are able to apply pest control measures in a timely manner.
3. Farmers do not have the burden of managing critical pest management operations, and have the time, therefore, to devote to other equally important operations.
4. Farmers are making a profit.
5. Consultants fill a critical niche in the industry, are successful entrepreneurs, and contribute to the state's economy though employment and income generated.
6. Technologically sound pest management operations practiced by farmers are helping to protect the environment.


Professionalism, people skills, and competence are traits a consultant should possess to be successful in the consulting business. Professionalism connotes a positive attitude, objective, unbiased assessment of the production needs of farmers, projecting a professional image, commitment to work, honesty, ensuring client confidentiality, inspiring loyalty among employees, and maintaining good relationships with extension. People skills include the ability to establish and maintain rapport, and to effectively communicate with farmers and peers, both verbally and in writing. Technical competence, constantly updated and based on the latest research, a sense of responsibility to provide timely and accurate information to farmer clients, a strong business sense, and a wealth of field experience accumulated over time are the underpinning of a successful consulting business.

## Sustainability and Replicability

The private cotton consulting business in Louisiana has increased steadily over the last 50 years and covers most of the cotton acreage and large cotton farms in the state. Return business over this period is evidence that private cotton consulting is a stable and sustainable business enterprise in the state.

Consulting is moving into other agricultural and horticultural enterprises. Consultants are now working in turf grass management, forestry, and the nursery business, and their presence is increasing. Currently, there are approximately 100 professional consultants in Louisiana. This number is likely to increase in the near future, with most of the expansion coming in the forestry and horticultural industry.

Expansion of the consulting industry in Louisiana will depend on several factors, including the needs of the forestry, turf and horticultural industries, increased pressure from environmental advocates, and greater use of consultants by other row crop producers.

The professional consultant offers a service that is of value in the fast-paced environment surrounding cotton producers. Producers have to deal with growing technological complexity, environmental stewardship issues, and regulatory controls. Professional consultants are in an excellent position to be of greater value in the future. The key is consultants who are well-educated and engage in continuously learning new technology. Improving education and preparing consultants for this role is the job of the land grant university system in the US. There is at least one university in the nation currently offering a doctoral degree in plant health. This type of degree may be in greater demand in the future. Universities will have to provide the impetus and support for this emerging discipline.

In the United States, consulting is still a young and growing profession. It began in the southern half of the United States and more specifically in the mid-south. It continues to spread to other areas of the country.

There is an opportunity for professional crop consulting in other parts of the world. Australia, New Zealand, and Western European countries have an established consulting tradition. Growth in these countries will be limited only by the availability of trained consultants and economic constraints. Developing countries of Asia, Africa, and Latin America, on the other hand, rely heavily on tax-supported, public sector extension to meet the information needs of farmers.

Except for specialized, large farms in these countries, it is unlikely that extension/information services will be developed in the private sector.

Elements of the private consulting industry that are replicable elsewhere include technical training and retraining, and licensing, certification, and regulation of consultants; a research-extension system to support consultants with new technology; a fee-based, contractual system offering a range of services demanded by farmers; environmentally responsible consultants and farmers; and a professional association for consultants.


## Lessons Learned

The basic economic principle of demand and supply is seen in this case. Farmer demand for services of value led to the creation of private, fee-based consulting services inasmuch as the university-based extension system did not have the personnel resources to meet this demand to the satisfaction of farmers.

Sustained, future demand will depend on how effectively private consulting services meet the economic and environmental goals of farmers.

Professional consulting has had positive economic, technological, and environmental impacts on row crop agriculture, particularly cotton. Increased acreage under contract and farmer adoption of consulting services, and higher economic returns have been obtained. Technological efficiencies have been demonstrated, and further degradation of soil and water avoided.

Expansion of professional consulting to other crop enterprises will hinge on farmer demand, technology, and availability of consultants.

The use of a professional consultant can enhance the management of a farm operation.

Training and education are of utmost importance in developing a competent professional consulting industry. The university system should be an integral part of training and supplying research-based information needed by professional consultants.

Private consultants need extension; extension needs consultants. Each fills a niche in the knowledge utilization system. Both need research. Private consultants rely on the extension-research system for up-to-date research-based information and training/education. Extension relies on consultants to provide specialized services that it cannot provide due to limited resources and its broader, more general education and service role in production agriculture. This kind of interdependency is a necessary condition for the introduction and growth of a viable consulting industry.

Consulting services can achieve broader environmental stewardship goals and regulatory provisions, and at the same time satisfy individual farmer needs. The development of consulting industries in the private sector in other nations will require the support and assistance of their education, research, and extension systems.

Key mechanisms initiating and guiding the change to private consulting services in this case include: (i) the teaching, research, and extension arms of the land grant university responsible for technology generation and utilization, and the training of consultants; (ii) technological, economic, and environmental drivers in the agricultural industry motivating farmers and business entrepreneurs; and (iii) licensing, certification, re-certification, and regulatory programmes for the private consulting industry under the auspices of the State Department of Agriculture.

Extension managers, including World Bank and other international staff, could be trained in these mechanisms through a series of seminars and workshops that would expose them to the underlying structure and process of the land grant university system, and the major concepts involved in initiating and developing private consulting services for farmers. In the United States, a state with a successful private consulting industry and its associated land grant university could organize this training. Other countries with similar systems and conditions could also provide this training. Interested countries and institutions could be invited to submit projects to the World Bank or other international organizations for funding consideration, depending on the need and scope of the training.

# Section Six

# OTHER CONTRACTUAL ARRANGEMENTS FOR EXTENSION SERVICES

# Chapter 16

# China: from Government-driven Group Contracting Extension to Farmers' Need-oriented Contracting Extension: the Case of Xinyang Prefecture in Henan Province

Yonggong Liu

## Introduction

Xinyang Prefecture is located in the southeastern part of Henan Province and the northern part of Dabie Mountain – a mountain in Middle China. There are eight counties and two municipalities under the administration of the prefecture. Six counties are designated poverty-stricken, located in the hilly region of the Dabie Mountain. Agriculture is important in the local economy. In the early 1990s, poverty of the rural population was a crucial constraining factor in its long-term development. Facing these problems, the governments at the various levels placed poverty alleviation and agricultural and rural development at the top of their development agendas.

Agricultural production growth in the prefecture stagnated after 1985, due to the remoteness of its mountainous geological location, the poverty of its rural population and the constraints existing in the agricultural research and extension services. To break through the stagnation, the prefecture government launched in 1988 a "Group Contracting Extension" campaign (GCE) for improving rice production.

As the name GCE implies, besides farmers and extension service centers, there was a group of contract partners involved in the group contracting extension services who played central roles in delivering the service to farmers' households. But the governmental bodies at various levels were in the leading position – acting as initiators, interest coordinators and main providers of initial funds of the GCE system. Since they assumed these key roles, these governmental groups dominated GCE.

## The Context for Introducing Contracting Extension

The "family responsibility system" was established in Xinyang Prefecture in the period 1980-1982, as in other provinces in China. After this rural reform, the former governmental extension service network at the township, village and production team levels was dissolved. At the same time, due to the market challenge farmers' demand for technology and extension service increased. During this period, the government extension service at county and upper levels was facing the challenge of reduced governmental funds. Under financial pressure, a free government extension service was no longer feasible. As a result, new hybrid rice varieties were not applied in the prefecture. According to local extension officials, before 1985 the sowing area of hybrid rice was only 10% of the total crops sowing area. Then, the GCE pattern was implemented in the prefecture in the period from 1988 to 1992 under the coordination and intervention of governments at the prefecture, county and township levels.

There were three sources of funds for financing the GCE model:

1. Governmental funds for agricultural development and poverty alleviation.
2. Short-term credit from the agricultural bank of China lent to the contracted households.
3. Farmers' payment to the qualified extension providers.

## The Contract

The contracting system in GCE is illustrated in Figure 16.1. There were three types of contracts in the system:

1. Administrative contracts were signed vertically between upper and lower level governmental rice production leading groups (RPLG), in which the upper level RPLG is the contract giver (responsible for formulating the contract), and the lower level RPLG is the subcontractor. In the contract, terms and quantitative targets were formulated by the upper levels and linked with rewards if the lower levels fulfilled the targets.
2. Technical contracts were also prepared by the RPLG at various levels and countersigned by technical groups, such as extension centres and research institutes at the same level. In this type of contract, research and extension tasks were outlined and the service performance was quantitatively linked with the governmental subsidy funds and rewards.
3. Service contracts were signed between extension centres, extension workers and farmers. In this type of contract, services were described in great detail. Terms included obligations of service providers, such as seeds, fertilizers, pest management, cultivation technology. Increased rice yield was the most important indicator for measuring the service's success. Obligations and

responsibilities of the service receivers (farmers) and payment rates and terms linked with agreed yield target were also clearly defined.

The broken lines with arrows (Figure 16.1) show the relationship between technical service supervision and instruction, technical contract groups at different levels, and farmers. The lines with two arrows indicates the contracting relation.

As mentioned above, in establishing and operating the contract system the government bodies played important coordination and supervision roles. Under the government's supervision there were also supporting institutions involved in the service system, such as the financial bureau, the Agricultural Bank of China, and the production company for delivering financial and materials service during the production period.

The GCE model, implemented from 1988 to 1992, made great contributions to introducing hybrid rice varieties and relevant cultivation technologies. After 1992, however, due to strong governmental interventions, violation of sub-contractors and the challenge of the market system, the GCE's effectiveness gradually deteriorated. Thus, since 1993, the GCE model has been gradually replaced by the farmer's need-oriented contracting extension model. In this model, extension centers and their workers and farmers are the major contract parties for special products and sectors, like cotton, vegetables, fruits and livestock products.

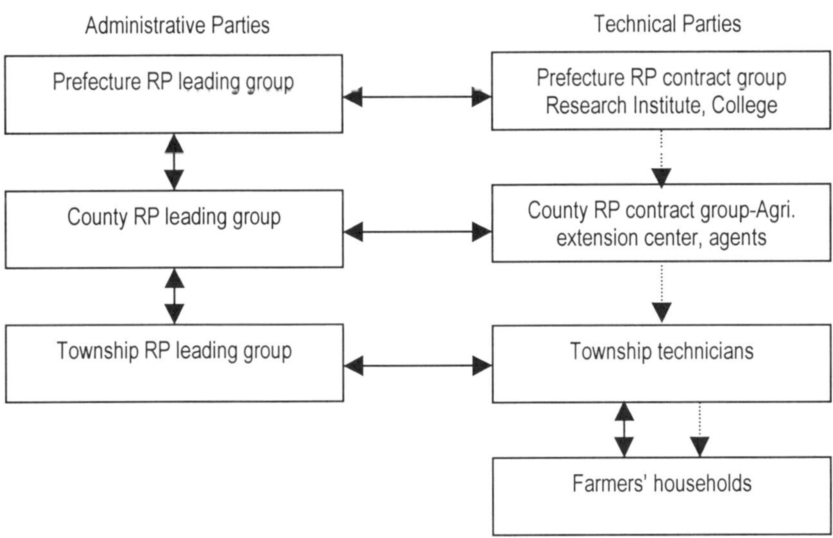


**Figure 16.1.** Organizational structure of GCE for rice production (RP) in Xinyang Prefecture.

## Impacts of GCE

Although GCE as a whole has been replaced by the new extension model, its long-term impacts can still be observed in the current production and extension service system. From 1984 to 1987 the rice production in Xinyang Prefecture stagnated. Hybrid rice varieties covered only 5-10% of the total rice sowing area. Through introducing GCE in 1988, the average yield per mu (1 mu = 0.67 ha) of hybrid rice varieties was 160 kg higher than the normal traditional varieties. Since its success in 1988, the total sowing areas of hybrid rice reached 1.05 million mu in 1989 and the average yield of hybrid variety reached 560 kg/mu, 230 kg higher than the traditional varieties. In 1990, the total contracted area was enlarged to 3.0 million mu, covering 70% of total sowing area. Today, the area of hybrid rice variety already accounted for more than 95% of total rice sowing area in the prefecture.

A first trial at contracting extension, GCE had a demonstrated impact on other sectors. From 1989 to 1991, the contract model was also transferred to other sectors and products, such as livestock production, tea, fruits, aquatic production, poultry, cotton, tobacco, etc. The GCE has put extension and users in a contractual relation and integrated multi-party incentives into a large contracting group. It provided conceptual experiences for establishing a farmer's need-oriented contract extension model in the prefecture after 1992.

Given governmental coordination and intervention in the contractual terms, farmers could get better technical service, such as seeds, fertilizers, machinery service, pest control, as well as technical training. At the same time, farmer's behavior toward the new technologies changed significantly. They also learned how to work with extension workers within commercial and contractual frameworks.

Through participating in GCE, technical service institutions – such as prefecture agricultural research institutes, agriculture colleges and county and township extension centers and stations – rediscovered their roles and mandates in the service. Since 1995 drawing on the experiences of GCE, these research and extension institutions have gradually changed their service mechanism from a traditional "top-down, instructional approach" into a "farmers need-oriented and service performance-oriented approach". This change has had long-term impacts on the establishment of the rural agricultural technology development and extension system.

A negative experience from GCE is that too strong governmental intervention during extension can constrain the farmer's and the technician's active participation and further create negative effects on the performance. This failure in GCE forced the governmental organizations to recognize the fact that in the market economy farmers should make their own decisions according to their social-economic conditions and the economic goal.

## Sustainability and Replicability of GCE Elements

Although GCE, as a complete model no longer exists, some basic considerations and elements are still useful for establishing an effective extension system in the market economy.

The most important aspect of the GCE is that economic incentives of all involved parties – farmers, extension workers and relevant governmental organizations – be sensitively reflected and integrated in the final contract. According to local extension officers, the GCE principles of "putting joint efforts, sharing the risks and benefits together with farmers" has become the basic concept of developing contracting extension models in the prefecture.

In current Chinese society, farmers are the weakest social group. Small farmers in the poor counties have very limited land resources, no access to credit and materials, less influence on the local market, and less power to control and mobilize the rural development resources. They are normally ignored by the governmental organizations and judged to be the most conservative social group in the social development process. Despite governmental intervention, the GCE has changed the traditional view of farmers by treating them as equal partners. This "farmer first" philosophy is extremely important for poverty eradication projects in other regions.

Compared with the former collective-extension system, GCE promoted the participation of farmers in the whole development process. As individual decisionmakers and subcontractors in the contracting system, farmers could have more influence from the beginning of the technology development and transfer process and they were able to evaluate the extension performance through the countersigned contract. In this context, GCE has helped empowered farmers to negotiate their interest and expectations with extension workers and other service institutions.

## Lessons Learned

For future reform of the Chinese agricultural research and extension system, the GCE case in Xinyang Prefecture can provide both positive and negative lessons; they are:

**1.** In the market economy, a customer need-oriented service is the general principle to follow for the service sectors. Agricultural extension as a technical service provider to large numbers of farmers must introduce the farmers' need-oriented extension service and an acceptable payment mechanism between extension workers and farmers. A contracting system under strong governmental intervention but without sufficient respect to farmers' needs and incentives can only reach the short-term goal, i.e. enlarging the sowing area of new varieties, but cannot be sustainable. Indeed, the governmental intervention in the Xinyang case had a negative impact on the sustainability of the GCE model. According to the extension officials interviewed at the prefecture level, farmers are quite sensitive to administrative and governmental intervention, but they need technical assistance from extension workers. How to redefine the governmental role in the individual contracting extension pattern is still an issue that needs to be clarified further. In China, the governmental organizations are still playing important roles in implementing agricultural policies at the grassroots level. In forming the social service system for agriculture, government can coordinate resources from different line agencies. There is no doubt that market need-oriented contracting extension still requires institutional coordination and policy support from the government at various levels.

**2.** As a trial in the transition from the former Chinese planned economy toward a more open market economy, GCE can be seen as a political extension campaign driven by government development goals. For disseminating certain technologies in a short time, it could still be effective to reach extension targets, for example, in poverty alleviation projects. However, in reality the more parties there are involved the more difficult it becomes to coordinate their contribution and balance their expectations from the service.

**3.** There were examples of violation of contract terms by some parties. Such violations could have led to collapse of the whole contracting system. This can also happen in the farmers' need-oriented contracting extension pattern. In order to avoid contract violations, effective and powerful supervision, monitoring and arbitration mechanisms would need to be established. The local government can play a role in arbitration and supervision.

**4.** Farmers' production associations or other relevant community-based organizations can act as group subcontractor signing the contract with the extension service so that the risks for each group member are reduced. This is also good for developing an inter-household cooperation mechanism within the group.

**5.** In the last 3 years the governmental funds for both the research and extension institutions have been reduced by more than 50%. Facing such challenges, the farmer's need-oriented contracting extension pattern can provide opportunities for research and extension institutions to change their government-subordinated status toward independent, market-based service enterprises. This tendency has been recognized by the central government and will be formulated in the reform policy for agricultural research and extension systems. But to prepare for such institutional reform, management staff of these institutions need to be qualified through training on enterprises and business management.


## Acknowledgement

In the preparation of this case, the author interviewed Mr Deng, Yanjun, Agronomist and Extension Officer of Agricultural Extension Center of Xinyang Prefecture (No. 30 Xinhua Xilu, Xinyang City, Henan Province, China. Post code: 464000, Tel./Fax: 0086 376 6380038). Mr Deng provided valuable information on the latest development of the extension models in Xinyang Prefecture. The author expresses acknowledgement for Mr Deng's assistance.

Chapter 17

# Finland: a Long History of Contracting for Agricultural Extension

Riikka Rajalahti* and Eija Pehu*

## Introduction

Two cases form the basis of this study: contracting by the national, although privately based extension provided by the Rural Advisory Centre, and contracting between agri-businesses, the university and growers.

## Finland 1-A: Rural Advisory Centres

### Case Identification

The first case refers to Finland's Rural Advisory Centres. The Rural Advisory Centre has entered into contracts with growers and other rural entrepreneurs for the past 100 years, primarily in the livestock and dairy sectors. Contractual arrangements in other agricultural sectors were carried out during the 1980s-1990s, and include growers interested in improving their economic planning, dairy and small livestock production, and rural entrepreneurship. In the future, the centre aims at developing contracts in still other sectors, such as crop production and horticulture.

***Editors' note:** The study was originally organized around three separate cases, including one on "contract farming" by agri-business, but the latter was excluded as not germane to the purposes of the present volume.

## Case History

Agricultural advisory work began in Finland in the late 1700s. In the 1860s, Finland had already established a model of development and advisory work based on farmers' initiatives (Niemelä, 1996). Interestingly, despite the strong support for the state organized rural advisory model in many countries, the arrangement of agricultural extension services did not become a function of the state in Finland. Instead, it remained the responsibility of private independent organizations. Although the agricultural policy leadership remained with the state and the advisory services received substantial financial support from the state, the state had very little decision making power over the type of services to be delivered. Thus, provincial organizations provided advisory services based on their own regional needs and popular initiative (Niemelä, 1996; Seppälä, 1999, personal communication).

The majority of the services were and still are delivered by voluntary farmers' organizations, country women's home economics associations, small-scale producers' associations, and other village-level organizations which together comprise 19 regional Rural Advisory Centres. The Association of Rural Advisory Centres and the Centre for Country Women and Homemakers are the Finnish-language umbrella organizations for the rural advisory bodies.

The information communicated by the Rural Advisory Centre after 1970, and its predecessor before that, was general in nature and concentrated on issues such as improved farming methods. The main clients for the services were private farmers. The services were mainly free, financed by public expenditure (Niemelä, 1996). By the 1980s, the economic situation in the rural areas had changed with strong emphasis on specialization, requiring specialized information as well as information on the management of complex farms. The challenge for the Rural Advisory Centre expanded as well-educated farmers increasingly sought information directly from research centres, specialized publications, and the private sector (Niemelä, 1996). General advisory services no longer met the requirements of a farm operating in a modern, more agribusiness-like economic situation. The services delivered had to be more demand-driven, which required the creation of service packages based on the producer's specialization.

Since the 1980s ecological farming and environmental issues have become part of the everyday concerns of the grower. During this time, information overload became a problem. Although growers became more educated and capable of looking for information themselves, the emergence of issues regarding ecological production, quality assurance systems, environmental impact, etc. resulted in an overflow of information. Consequently, growers had difficulties in choosing the information relevant to their particular situation.

In general, the number of farm families began to dwindle and an increasing number of farms derived most of their income from activities other than farming. The agricultural advisory services no longer met the needs of these farms, and therefore, extension broadened its programmes to include services suitable to various rural entrepreneurs. Subsequently in 1990, the agricultural extension services were recast as rural extension services (Niemelä, 1996).

## Impacts

Contracting resulted in changes in the funding and delivery of services. Prior to the 1990s, most of the activities carried out by the Rural Advisory Centres were financed by public expenditure and membership fees. Staffing in the advisory service depended first and foremost on the level of state funding, since salaries were the organizations' largest expenditure and state aid their most important source of income (Niemelä, 1996). In 1995, state subsidies were negotiable based on the activities and the development of the advisory services. Since then, the allotment of state funds for general extension services has decreased (by more than 40%) whereas customer service fees have increased (Anon., 1996).

At present, the operations of the Rural Advisory Centre, which are controlled by the rural entrepreneurs and other members of the Rural Advisory Organization, are primarily financed by customer service fees (about 50%) and state rural development subsidies (20-30%) as well as proceeds from other activities, such as special project funding and extension materials. Contracting has resulted in part from the decline in state funding. Since Finland joined the European Union (EU), the need for advisory services has increased tremendously. Simultaneously, the centres have also received a new source of funding. EU funding has made it possible to provide support for general advisory work as well as the development of new services such as quality control systems (Anon., 1996; Seppälä, 1999, personal communication).

The Rural Advisory Centres still provide advisory services, but in general they do not have the personnel nor financial resources for general extension work (Seppälä, 1999, personal communication). Therefore growers themselves have had to assume responsibility for finding information previously provided by the Rural Advisory Centres. To help meet the demand for agricultural information, a few Web-based (Internet) advisory services have been created in collaboration with the Rural Advisory Centres, the Ministry of Agriculture and Forestry, agri-businesses as well as public institutions (Anon., 1996; Anon., 1999). Thus, the Centre operates as a consulting service delivering comprehensive advice on the management of farm enterprises.

Contracting for personal delivery of services has become extremely important. The Rural Advisory Centres provide various development services, such as appraisal of a business idea, entrepreneurial training, planning of operations and production, aspects of business economy, taxation, and marketing. Assistance is also provided to help in the development of quality systems for farms and small-scale enterprises (Anon., 1999; Seppälä, 1999, personal communication). The extension services and materials, along with research and training, are customized to meet the needs of farmers and other rural entrepreneurs. Comprehensive training courses, seminars and field days introduce the latest research results and applications to farmers (Anon., 1999).

Rural Advisory Centres are moving towards signing long-term contracts with their clients (Seppälä, 1999, personal communication). These contracts are always case-specific and carefully state what services at what price shall be delivered. For example, signing a livestock development package contract with a Rural Advisory Centre, often for more than 10 years, gives many advantages to the farmer. The investment in the contract generally pays for itself in the first year. Producers can choose from a variety of services, such as expert planning and monitoring of systems aimed at cutting down the costs of feed and fodder production, upgrading animal progeny, improving the quality of the products and developing efficient working methods according to the specific needs of the enterprise (Anon., 1999). Previously, the advisory services consisted mainly of planning activities; since the 1980s, however, an important part of the service package has been to monitor and follow-up the outcome of farm-specific action plans.

## Critical Factors for the Success of Extension Contracts

For several decades contracting in Finland has been successful. The primary reason for this success is the overall benefit to all parties involved. Contracts between the Rural Advisory Centre and rural entrepreneurs have enabled the Centres to better assess the needs of their diverse clientele, thereby allowing them to concentrate their resources on the services most needed. Contracting has also contributed to the increase and diversification of clientele.

Clients receive and pay only for the information they need to run their business; thus customer satisfaction has increased as a result of contracting. An important factor in the planning of the development and content of the services is the participation of the clients (members of the associations). Regular monitoring and follow-up of the action plans enable the use of indicators, which vary according to the services provided – e.g. production figures (livestock, crops, fisheries), ratio of inputs to outputs (all sectors), health (e.g. use of antibiotics in livestock, pest incidence in crop production), and in the case of environment, nitrate and pesticide residue levels. It can be concluded that the Rural Advisory Centre now has a sound basis for its contractual extension services. Its present ambition is to widen the services provided more towards crop production, quality systems and environmental management. For the Rural Advisory Centres, the overall success of production and the quality of the produce serve as indicators.

# Finland 1-B: University

## Case Identification

The Faculty of Agriculture and Forestry of the University of Helsinki, the sole institution of higher education in agriculture, does not have a mandate for agricultural extension and thus traditionally has not been involved with rural advisory services in Finland. However, during the past decade its approach has changed somewhat due to the Integrated Production (IP) programme between the University, industry and growers. Other sectoral public research institutions have also started to provide services on a commercial basis for the implementation of research as well as for providing crop protection advice and weather forecasts.

## Case History

The Faculty of Agriculture and Forestry has the national mandate for higher education in agricultural sciences. Traditionally, it has not been involved with agricultural extension aside from minor *ad hoc* collaborations with growers. In addition to the University, which operates under the Ministry of Education, the Agricultural Research Centre (ARC), which operates under the Ministry of Agriculture, has also become involved in agricultural extension.

The ARC and the University have by and large operated in isolation from each other and from the activities of the Rural Advisory Centre, aside from occasional cooperation in research activities. Due to lack of participation of growers in the generation of research questions and planning of research projects, criticism grew in the 1990s over the relevance of the research conducted in public institutions. This situation provided the University with an open field, but with no predetermined ways of pursuing collaboration with growers.

## Impacts

The introduction of farming systems research methodology into the Department of Plant Production of the University of Helsinki by Professor Eija Pehu in the beginning of the 1990s turned the attention of the research staff towards the concerns of the growers. This shift was further strengthened when Finland became a member of the EU in 1995. More competitive agricultural production systems suited to the severe agro-ecological conditions of Finland were called for. Research conducted in the University and in the sectoral research institutes gradually grew closer to the growers and the growers' perspectives were integrated more fully into the research programmes. Public research institutions thus became part of the rural extension system. Their research findings are increasingly shared with and even sold to the growers who recognize the value in knowledge-intensive farm production systems.

The IP Programme of the Department of Plant Production of the University of Helsinki serves as a good example of the cooperation between research and farm production. While the main focus of the programme was to develop a sustainable, economically feasible IP system for cereal-producing farms in southern Finland, early on the programme also realized the need for participatory, customized expert services for the farmers. To test the concept of participatory extension, 11 pilot farms were selected from a group of interested growers. At the onset of the project each farm was visited by the research staff, crop rotation plans were made for each individual farm (Rajalahti *et al.*, 1998; Riekki *et al.*, 1998), and a cooperation contract signed. The contract required the researchers to provide advisory services, arrange seminars and training events, and to cover the expenses of soil analyses. The growers, in turn, agreed to follow IP-farming methods, maintain records of all farming activities, collect data as specified in the contract, and allow farm visits (Rajalahti *et al.*, 1998; Riekki *et al.*, 1998). A contract was also signed between a leading food processing company and the university since the company needed guidance in the formulation of their own IP-guidelines for the growers. Thus, the university IP research team arrived at providing advisory services and planning inputs for applied research and in turn received funding for the IP research programme (Pehu, 2000, personal communication).

Another example involving the Department of Plant Production is the contractual arrangements between growers and a research team interested in introducing a new oilseed crop into Finland, *Camelina sativa*. Successful research efforts along with cooperative efforts with farmers resulted in the formation of a company, which now has a production contract with the original participating farmers and is also negotiating shareholder arrangements with contract farmers (Alén *et al.*, 1998).

**Critical Factors for Success**

In the case of the university IP Programme, fundamental changes occurred prior to the collaboration between the private sector and pilot farms with the university. First, as the importance of environmental issues emerged, farming systems research gained in importance. Research station-based farming systems research cannot, however, be continued reliably without testing and verifying the methods and results on active farms. Thus, the participation of pilot farms was a necessity for the continuation of the IP-research (Rajalahti *et al.*, 1998). Second, the consumer demand for more environmentally friendly products made the production of IP and ecologically produced food more profitable. Third, it became more acceptable for a public institution to receive funding for research from the private sector. A win-win situation between the pilot farms and the university IP project promoted the participation of both parties in the active development of IP systems for Finnish growers. Growers in turn received the latest information on new production methods, which enabled them to gain a market advantage and to sell their produce at a higher price (IP-trade mark). Information gained by grower participation proved to be an invaluable source for improvements in the development of IP farming systems.

## Concluding Remarks

National agricultural services and later rural extension services have always been managed by private organizations in Finland. In turn, the extension activities of private organizations have been financed by the public sector, membership fees and specific project funding. The proportional shares of these individual funding sources have changed over the past decade with the public funds being considerably reduced as a new player, the EU, has been introduced, affecting structural development.

Supplemental to the national extension service network, several agri-businesses have contractual arrangements, which include advisory services, with producers. Recently, also university research programmes and sectoral research institutes have reached out to collaborate with the farmers. The content of the services has moved from the traditional agricultural extension recommendation packages to customized, commodity specific, entrepreneurial support services. These structural and programmatic developments have resulted in national agricultural policy framework changes, such as the reduction in agricultural subsidies, increased attention to the environmental impacts of agriculture, development of quality assurance systems, institutionalization of ecological production and marketing systems and support to small and medium size entrepreneurs in agro-processing. These changes pose new challenges for the rural extension services, which have already responded and are widening their knowledge base and the scope of services they provide.

## References

Alén, K., Kallela, M. and Pehu, E. (1998) Alternative oil-seed crop *Camelina sativa.* Contract-No. AIR3-CT94-2178. Periodic Progress Report, Finland (1.1.1998–31.12.1998).
Anon. (1996) *Maaseutuneuvonta 1995.* Maaseutukeskusten Liiton julkaisuja No. 908. ISSN 0789-9661. SPOY Kokemäki, Finland.
Anon. (1999) Rural Advisory Centre. http://www.agronet.fi/mkl/yleiset/inenglis.htm
Niemelä, J. (1996) *From Provincial Shepherds to Rural Advisory Centres.* Finnish Historical Society and Association of Rural Advisory Centres, 519 pp.

Rajalahti, R., Riekki, A., Loiva, M., Poutala, T. and Pehu, E. (1998) Optimized rotation and more efficient utilization of natural resources: experience from pilot farms in Northern Europe. *Proceedings of the 15th International Symposium of the Association for Farming Systems Research-Extension.* Pretoria, South Africa, pp. 267-271.

Riekki, A., Loiva, M., Pehu, E. and Poutala, T. (1998) Integroidun peltokasvituotannon pilottitilaprojekti. Loppuraportti, 36 pp.

## Contacts

Mr Juha Seppälä, The Rural Advisory Centre
Email: Juha.Seppala@agronet.fi

Mr Timo Kaila, Lännen Tehtaat Ltd., P.O. Box 160, 27821 Iso-Vimma, Säkylä, Finland
Phone: +358-2-8397 4600
Fax: +358-2-8397 4622

Ms Karita Alen, Camelina Ltd.
Phone: +358-9-191
Fax: +358-9-191
Email: karita.alen@helsinki.fi

Chapter 18

# Mozambique: Dual Public-Private Services for Small Farmers

Hélder Gêmo and William M. Rivera

## The Setting

The Republic of Mozambique on the southeast coast of Africa, a country almost twice the size of California, is organized into ten provinces. Nampula and Zambezia located in the north are the provinces where the Government of Mozambique (GOM) intends to initiate the first two projects in its move toward developing a pluralistic national system of rural extension. This move is part of the initiatives planned by the GOM following the loan it negotiated from the World Bank in 1998 to improve its agriculture sector.

## First Phase: Late Beginning

Mozambique's public-sector rural extension service is one of the most recent in the Southern Africa Region as well as in sub-Saharan Africa. Not institutionalized until March 1987; its initial development took place between 1989 and 1992 (Amisse, 1997; Gêmo, 2000). At that time, state farms contributed considerably to the formal establishment and initial consolidation of public-sector rural extension.

The state farms basically provided field technicians with an elementary academic level in agriculture and professional field experience, especially in cash crops such as cotton, cashew and to a lesser extent in tobacco. Government organizations (GOs) as well as the majority of international non-governmental organizations (NGOs) were involved at that time in emergency activities relating to drought and civil war.

The international NGOs contributed significantly to the growth of public extension during this initial phase of extension's development. Ibis, for example, the Danish Association for International Cooperation (a Danish NGO), was fully integrated with public-sector extension in Zambezia province (1992-1998), providing institutional support. The German Society for Technical Cooperation (GTZ, a government organization) had a close relationship with public-sector extension in Manica province (1989-1999), giving technical assistance and some institutional support.

The UN agencies such as FAO, UNICEF's integrated rural development programme and UNDP made valuable contributions to technical and financial support. The International Fund for Agricultural Development (IFAD) also played a significant role as one of the first (and still current) financiers of public-sector extension.

During this first phase of extension development, the environmental situation was less than ideal. In addition to the prolonged drought in the country, a high degree of insecurity existed in the rural areas as a result of civil war. There was a hasty exodus of rural families seeking relatively safer zones. Thus too, new extension networks had a tendency to be established near to the provincial capitals or in zones with some level of safety. Even in those zones, however, the daily tasks of extension workers were completed in an average of 4-5 hours in order to avoid dangerous early morning and afternoon hours. Crucial activities such as supervision, monitoring and evaluation (M&E), on-the-job and in-service training were also affected by these adverse circumstances. In short, normal extension activities were not undertaken until after the civil war, 1992-1995.

## Second Phase: PROAGRI

In 1998, the GOM initiated the World Bank-financed National Agricultural Development Programme (PROAGRI). Public-sector rural extension is one of the eight components that comprise PROAGRI.

In line with the general objectives and guidelines of PROAGRI, the GOM developed a National Extension Master Plan (1999-2003) that called for the development of an integrated national extension system, SISNE (Sistema Nacional de Extensão Rural). The Ministry of Agriculture and Rural Development (MADER), through its Department of Agricultural Extension (DNER), proposes to collaborate with other organizations interested to provide extension, such as NGOs, farmers' organizations, private-sector commercial farmers, registered groups of certified extension workers, and other private-sector entities. Thus, within the framework of PROAGRI, DNER will undertake two major tasks: the operation of a number of public-sector extension networks throughout the country and the management of private-sector providers, especially in areas where there are no public extension networks.

# Current Phase: Outsourcing

The institutional pluralism of extension service providers is considered to be a major strategy in the development of the National Extension Master Plan (1999-2003) and the advancement of an integrated national extension system (SISNE). The Extension Master Plan states that publicly financed extension is open to multiple financial and delivery arrangements. These arrangements include outsourcing, cost sharing with private and community extension structures, and cost-recovery initiatives with individual farmers and farmer's groups and associations.

Outsourcing is the principal avenue selected for involving private-sector providers, some of whom are already operating extension systems parallel to that of the public sector. GOM plans to outsource the delivery of extension in two pilot projects in two districts, one in Nampula and one in Zambezia. As defined in this instance, outsourcing means "contracting out" responsibility for extension delivery to private-sector providers (Rivera, 2000). It is the act by public-sector extension to promote and support private-sector involvement in extension provision (whether by private companies, NGOs, farmer associations, or registered individual extension consultants). At present, one of the objectives of this outsourcing initiative is to prepare DNER to coordinate, oversee and regulate private-sector providers and to learn from the experience of outsourcing.

# Problems Confronting the Development of SISNE

### Young Institutions

Outsourcing extension delivery services to private-sector providers is the initial step being taken to develop an integrated national extension system (SISNE). Within that context, the GOM has decided first to open a bidding process directly toward national NGOs. It is expected that NGOs will best serve the initial purposes of the outsourcing project in part because they are likely to create a demand-led system which is responsive and accountable to farmers' agricultural problems and goals, as well as being operationally effective and financially efficient.

The task of providing extension education to farmers is, however, enormous – in part because of the country's large size, the lack of infrastructure mentioned earlier, and the national urgency to promote increased and sustainable agricultural production. The involvement of both public-sector and private-sector extension providers is seen as complementary and strategically crucial if a nationwide system of extension is to be developed. Notwithstanding, the private-sector extension NGOs focusing on smallholders on a non-profit basis are in the initial stages of development, as is the public sector.

## Lack of Collaboration Among Extension Providers

To date, there are few instances of collaboration between public and private (non-profit and for-profit) extension entities. Public-sector extension had (and still has) some notable partnerships with international NGOs, such as Ibis, CLUSA (the Cooperative League of the USA), and Sasakawa Global 2000.

Apart from the above, there are no significant examples of collaborative relationships between public-sector extension and non-profit private extension providers. In general, even collaboration between NGOs is weak, even when they are both working within the same province. NGOs tend to be small in size, independent, and differ in their philosophies; these attributes, as Farrington (1997) notes, militate against the creation of effective forums, whether at national or provincial levels. In the last 2-3 years some forms of informal and occasional relationship have started to grow between public services and NGOs, particularly at the provincial level, with some national and international NGOs. With few exceptions, the same cannot be said in relation to private enterprises doing commodity extension.

## Lack of Farmer Organizations Providing Extension

The Government of Mozambique hopes and expects to expand the present pluralistic approach to extension by eventually being able to include farmer organizations in the SISNE plan. The promotion of farmer organizations currently constitutes a task for both government and private-sector extension, although for different reasons depending on whether the private-sector entity is a for-profit or non-profit organization. In Mozambique there are very few known examples of farmer organizations carrying out extension or supporting such services in rural areas. In general, these types of organizations are in an initial stage of development. But, they are seen as having considerable potential for promoting farmers' empowerment and local sustainability.

## The Joint Venture Companies

Joint Venture Companies (JVCs) carry out purely commodity extension (e.g. for cotton, tobacco, and cashew). It is irrational to try and engage these organizations in the private sector to undertake projects involving diversification, unless a profit can be clearly anticipated. Also, their orientation is quite different from that of government or the NGOs. JVCs tend to consider external inputs as the routine mode for production of cash crops, while Government encourages the rationale of chemical inputs by farmers, highlighting environmental and human, not only economic, issues. However, the outsourcing project envisions the involvement and contribution of JVCs, probably under NGO subcontracting to deliver input supplies and/or develop markets.

## The Challenge

The present initiative towards an institutionally pluralistic national extension system is confronted with significant external problems, some of which are outlined in the previous section. In addition, public-sector extension must confront two internal challenges, having to do with organizational structure and management. One challenge is to review and clarify the goals of its public-sector delivery service, given the proposed development of private-sector delivery services. The second challenge is to develop its management capability both as a provider of extension services and as coordinator, overseer and regulator of private-sector extension delivery services.

In the year 2000, public-sector extension was comprised of approximately 700 agents, of those only about 200 are permanent employees; the rest serve on a contractual basis. Many questions arise. Should the number of permanent employees be increased? What should be the status of the contracted agents? If the public sector service is to be expanded, what will be its goals? How can those goals best be accomplished? Does public-sector extension have a professionally adequate number of administrative and field staff to realize its goals?

In addition to reviewing its priorities and within that framework setting realistic goals to be accomplished, public-sector extension will need to develop a systematic and useful (and effectively used) management information system founded on regular monitoring and evaluation. The latter performance and programme control features become particularly important as public-sector extension moves toward a bifurcated system in which it must both operate publicly provided extension services and coordinate, supervise and regulate privately provided extension services.

## Next Steps

The outcome of the outsourcing project depends on several factors. On the positive side, extension management staff, from top to bottom, are comprised in general of young professionals who are willing to innovate and try new forms of organization and intervention.

There is notable commitment and a growing understanding among central extension management staff that outsourcing is neither an alternative to public-sector extension nor an effort to prove that public-sector extension is ineffective. Outsourcing is beginning to be seen as a way to develop a national system of extension delivery, to promote the participation of other actors and further pursue the involvement of farmers in extension programmes. The view of the role of government is changing, from the only provider to one among many providers of extension, functioning also as regulator, supervisor, evaluator and facilitator of extension provision. This changing view among management staff is important because it creates a favourable and motivated work environment.

On the negative side, educational levels of extension staff are low. At the provincial level (with the exception of Nampula where there are five technicians with BSc degrees) only one or two technicians have BSc degrees. Thus, training of technicians and/or the recruitment of new qualified staff will be necessary for monitoring and evaluation, as well as in assisting the pilots. DNER at the central level will have a crucial task in guiding the process of outsourcing and developing and training provincial staff in supervision and M&E. Outsourcing demands systematic M&E. Institutional development involving human resource advancement in new knowledge and skills is also needed urgently.

Mozambique's move toward a pluralistic national rural extension system has implications for the nation's agriculture, and also for its social and economic development. The outcome of Mozambique's move may also have an impact on the determination and direction taken by other developing countries confronted by the exigencies of poverty and food insecurity and the demanding challenges of globalization.

# References


Amisse, J. (1997) A importância da extensão rural. In: *A Extensão Agrária em Moçambique (1987–1997)*. Direcção Nacional de Extensão Rural, Maputo.

Farrington, J. (1997) The role of non-governmental organisations in extension, (Chapter 23). In: Swanson, B.E., Bentz, R.P. and Sofranko, A.J. (eds) *Improving Agricultural Extension: a Reference Manual*. Food and Agriculture Organisation of the United Nations, Rome.

Gêmo, H. (2000) Breve resumo histórico da extensão pública em Moçambique. In: *Extensão Rural em Moçambique*, 1:1, Janeiro 2000. Direcção Nacional de Extensão Rural, Maputo.

Rivera, W.M. (2000) Manual for outsourcing extension. Draft. Direcção Nacional de Extensão Rural, Maputo and The World Bank, Washington DC.

Chapter 19

# Uganda: Private Sector Secondment of Government Extension Agents

L. Van Crowder and Jon Anderson

## Introduction

This country case study contrasts theoretical and practical experiences in contracting of extension services in Uganda. The plans of the Government of Uganda (GoU) for the agriculture sector call for increased decentralization, in line with overall policy, and for contracting out extension services. The longer term national planning ideal is a commercialized agriculture sector dominated by the private sector and supported by private advisory services. Agriculture planning calls for a decreasing role for public extension with a corresponding increase in private extension provision. However, the local field reality at present is somewhat different. Concrete examples of contracting out extension upon which to build policy and plans are scarce. In contrast, there are interesting, local level practical examples of public-private extension coalitions or partnerships. Within these coalitions, non-government partners buy into the government system for technical services. These coalitions seem to present complementarities and synergies that merit further analysis and may provide lessons for consideration in extension planning and policy.

## Background

A basic component of the GoU's strategy for more relevant and responsive extension is the contracting out of field extension services (see Modernization Plan for Agriculture). It is envisioned that public-sector funds will be used to contract non-public providers of extension services. Possible providers are NGOs, farmer associations, private commercial firms and teams of agricultural technicians among others. The interest of government and donors is to shift over time from public-sector funding and delivery of extension to public-sector funding and private delivery, with the ultimate goal of achieving both funding and delivery of extension by the private sector in about 20 years.

Currently, the provision and management of extension services is the responsibility of district-level governments, although certain support functions are retained at the central level, including planning and policy formulation, regulatory functions, technical backstopping and training, designing extension methods and setting standards for extension practice. In 1998, the Directorate of Extension in the Ministry of Agriculture, Animal Industry and Fisheries (MAAIF) was abolished, central staffing was reduced by some 80% and the major responsibility for supporting field-level extension was transferred to the National Agricultural Research Organization (NARO).

District extension staff salaries are paid by unconditional block grants from central government. Under the decentralizationn plan, any new extension staff recruited to work in a district are to be paid by local government. Districts pay operational expenses for extension for the most part, although sub-counties are expected to contribute to extension costs. Extension field activities are greatly restricted by operational funding, which falls far short of what is required. Donors and non-governmental organizations (NGOs) often offset shortfalls in district funds, utilizing district extension staff to implement their project activities, providing operational funds and often a salary supplement.

In a move that seems contrary to downsizing and restructuring objectives, central government announced plans in 1998 to employ "Graduate Specialists" as new district extension agents, with up to three graduates deployed to each sub-county. Salaries are budgeted by central government in the form of conditional grants, while districts and sub-counties are expected to provide the bulk of the operating expenses. The national goal initially is to hire 800 Graduates Specialists. If three graduates were eventually hired for each sub-county, the result would be approximately 2700 new public-sector extension staff.

Not surprisingly, donors are concerned that the hiring of Graduate Specialists as civil servant extension agents is contrary to the GoU strategy outlined in the modernization plan to increasingly contract out extension services to the private sector. The hiring of additional civil servants is *not* viewed as contributing to an extension system that is more operationally efficient, cost-effective and accountable to farmers – characteristics, supposedly, of private-sector contracting arrangements.


## Contracting Extension: Variations and Options

At present, there are very few cases of the public sector funding private-sector delivery of extension in Uganda. There are, however, various examples of the reverse – private-sector funding and public-sector delivery. Before looking at one of these examples, it is useful to consider the contracting approaches being explored by government and donors supporting extension.

The extension contracting option that has received the most attention by the GoU and donors in the design of a national extension programme involves the provision of conditional matching grants to local governments to contract extension services. Matching funds from local governments would initially be small (5-10%) with the goal of increasing the amount over time as well as applying cost sharing measures with farmers, NGOs and others. Grants would be provided to district and sub-county governments to contract extension services, with the emphasis on provision of field extension at the sub-county level with technical backstopping from the district.

Farmer participation would be through Extension Advisory Boards (EABs) or similar entities. The EABs would have majority representation by farmers, but might also include elected officials and local government civil servants as well as one or more extension agents assigned to the sub-county. Decision-making regarding extension work plans and selection of service providers would rest primarily with the farmer representatives and the elected local government officials. Monitoring and evaluation of contracted extension services would be done by the EABs, which would certify payments to service providers according to the terms of the contract and in accordance with local government regulations.

It is envisioned that in some cases, sub-counties might opt for an arrangement with a farmers' association, such as Uganda National Farmers' Association (UNFA) or the Uganda Co-operative Alliance (UCA), both of which have experience in managing and providing extension services, including on a fee basis. In this approach, EABs would approve a proposed extension plan of work and a service provider would be selected by the farmers' association and the EAB, or the farmers' association would be the service provider itself using contracted extensionists.

In the first instance, the sub-county would contract with the farmers' association to manage a sub-contracted entity, such as a local NGO or private team of extensionists, to provide extension services to local farmers. In the second instance, where a farmers' association has a well established membership, the sub-county would contract the farmers' association to provide extension services to its members as well as to other organized groups of farmers who would pay a fee. Salary and operational costs would be covered by the conditional grants and cost-sharing provisions would be applied.

Provisions would be made to strengthen the ability of sub-counties and the EAB to prepare local extension plans and contracts for extension services and to manage and evaluate contracted services. Checklists for preparing contracts, model bidding procedures, sample contracts, inventories of extension providers, qualification criteria for providers and guidelines for evaluating and selecting providers would be prepared. Field testing of contracted out extension arrangements would take place in order to evaluate the cost-effectiveness of the various arrangements.

This approach conforms to the anticipated benefits often attributed to contracted extension: more operationally efficient and cost-effective extension; extensionists that are more accountable for their performance and results; and a diversity or plurality of service providers.

In contrast to the above proposed contracting options, currently in Uganda the most common extension contracting arrangement is the "secondment" of district extension staff to NGOs. In many districts, there is a pool of extensionists who because of a lack of operational funds are under-utilized. NGOs and donor-supported projects often lack technical staff partly because of government monopoly of technical expertise, for example in agronomy, soil science, forestry and other disciplines. This is because government provided technical training for free and guaranteed (or previously guaranteed) the employment of agricultural graduates. Essentially all technicians ended up with the government service. This approach has largely disappeared, but the non-public sector has yet to emerge with much technical capability although this is beginning to change with government downsizing and increased non-public employment opportunities. NGOs and donor supported projects thus draw on the public staff pool to contract extension workers, providing operational funds, travel allowances, per diem and in some cases salary supplements to augment low civil servant wages. We term this arrangement "contracting in" extension.


## "Contracting In"

There are private/public extension arrangements presently operating in Uganda which involve public-sector extensionists in contractual arrangements with the private sector with the goal of identifying and actualising mutual benefits. One such example involves a bilateral donor, an international NGO, national NGOs and public-sector extensionists. In this example, the international NGO is contracted by the donor to operate an agricultural development programme involving a grant management component. Grants are provided to locally based international NGOs and national NGOs to implement field activities focused on increasing production of basic food products. These NGOs contract public-sector extension staff to extend improved farming techniques to groups of farmers.

The international NGO is ACDI/VOCA, which operates a PL 480 Programme funded by USAID. Through a Food Security Fund, grants are provided to national NGOs to carry out extension activities focused on increasing production of four crops – maize, beans, cassava and oilseeds. In 1998, 18 grantees were funded to work in 37 districts. Thirteen of these grantees were Ugandan NGOs and five international NGOs.

Appropriate Technology Uganda (ATU), a local branch of an international NGO, was a grantee of the ACDI/VOCA grants programme. ATU utilized district-based extension staff to carry out the field demonstrations that are part of its project activities. In 1999, ATU operated grant-funded projects in seven districts, with activities in seven sub-counties in each district for a total coverage of 49 sub-counties. An appointment letter was used in "contracting" district extension staff as Farmer Participatory Research Assistants (FPRAs) for specific field activities. They operated under an FPR Supervisor contracted full-time by ATU. The terms of the appointment for the FPRAs included:

**1.** Residential training on farmer participatory approaches and improved production practices.
**2.** Monthly planning meetings to draw up work plans and targets, for which FPRAs were reimbursed for travel costs.
**3.** Payment to FPRA each month of up to 8 days of "safari" (travel) day allowances depending upon the actual work plan and activities carried out.
**4.** FPRAs were expected to supervise planting material multiplication and distribution, conduct farmer training, supervise on-farm trials and demonstration plots and organize field days.
**5.** FPRAs were required to submit a monthly field report on activities carried out and achievements against impact targets.
**6.** Payments of allowances and transportation refunds were contingent upon submission of satisfactory monthly reports, which were approved and checked for reliability by the FPR District Coordinator.
**7.** ATU supplied FPRAs with a new bicycle, which becomes the property of the FPRA after completing 3 years with the project.
**8.** The appointment was conditional upon the staff person remaining an employee of the Public Sector District Agricultural Service.

In contrast to situations where government "contracts out" extension to non-public providers, the Uganda case, where non-public entities contracted with the government for the provision of services, is an example of "contracting in" extension. District extension staff, who are often inactive because of a lack of operational funds, were mobilized by the NGO to deliver what appears to be effective technical extension services targeted to small-scale producers of food crops. The technical agronomic knowledge of public extension agents, a unique resource in a country where trained human resources in agriculture are lacking, was utilized in a situation that seemed to provide benefits to the public sector, the NGO and to farmers.

## Impacts

The arrangement affected the institutional partners as well as the farmers. The skills of extension agents were updated by training in improved production practices and they learned how to foster the participation of farmers in a range of activities (e.g. on-farm trials and demonstrations, planting material multiplication and distribution, field days). The skills that the NGO had in planning, organizing and supervising local development activities, including group promotion and participatory, community-based actions, was combined with the technical skills of extension agents, who in many cases have in-depth knowledge of local farming systems. The NGO and the farming community appeared to have benefited from the technical skills of the extension agent. The arrangement also has the potential for contracted-in extension agents to provide a mechanism to counter balance the NGO agenda and perhaps integrate public good types of concerns (e.g. environmental protection, HIV/AIDS) into project activities. Direct quantifiable impacts on farmer's production, productivity and income from this type of arrangement have not been studied as of yet, but positive mutual benefits can be expected (e.g. improved cost effectiveness as a result of public-private cost sharing of extension).

## Sustainability and Replicability of the Contract

In Uganda, the public-sector in effect was cost-sharing the expense of the extension services since the salaries of the contracted extensionists were paid by district funds. The cost-sharing agreement did not involve the actual transfer of limited public financial resources. Rather, the agreement was based upon complementary functions and expertise, with the NGO benefiting from a pool of already salaried agricultural technicians and public extension benefiting from the operational funds, training and a supervised plan of work provided by the NGO. The incentives for public extension agents to participate were high (e.g. allowances and a bicycle), and not surprisingly there was considerable competition among agents for contracts. Good performance was ensured since the NGO could withhold allowances and dismiss contracted extension agents if they did not fulfill work expectations. It is reasonable to expect that extension agent motivation, and skills obtained in farmer participatory approaches, resulted in benefits for farmers.

Since the participation of district extension staff was conditional on their remaining public-sector employees, the net effect is not a reduction of the public cost for extension staffing, although the operational burden of extension is alleviated in those districts and sub-counties where project activities are implemented. In fact, such an approach may increase the demand and justification for public sector staff. The Uganda coalition appears to be a "win-win" situation – the NGO did not have to hire personnel and got inputs of scarce technical expertise and government extension got increased mobility, effectiveness and training for its staff. The synergy and complementarity of functions resulting from this private-public extension arrangement may be more efficient and effective than either sector working in isolation, which would require each organization to have the full range of functions. This could result in duplication of functions, or implementation of conflicting extension activities. Because of the fluid and location specific nature of these coalitions, replicability in terms of using this experience as a model might be difficult. The coalition to a large extent depended on local networks, available staff skills and organizational needs; its sustainability would require the building of relationships among the NGO, the public extension sector and project farmers that result in reciprocal benefits. It is worth noting that there other examples in Uganda of NGOs contracting in public extension provision and the authors have anecdotal information about similar arrangements in other African countries.


## Lessons Learned – Recommendations for Improvement

The contracting in arrangement described above is not a perfect one, and there is clearly scope to improve it. Concerns can be raised about farmer involvement in developing the extension programme of work and evaluating the performance of contracted-in extensionists, although the provision of community participation in project development and implementation is used by ACDI/VOCA as a criterion for assessing grant proposals. Also, ATU claims that it promotes farmer participatory approaches and involves representative groups of farmers in planning extension activities.

There is also the problem that the ATU extension programme was not developed in the context of district agricultural development plans – the fact that district plans often are poorly prepared and under-supported means that the donor and NGO can rather easily set the agenda. On the other hand, the NGO claims that it seeks district extension and local government approval of its extension programme of work and of course used district extension staff.

It might also be argued that public extension agents who are under agreement with an NGO are in a situation where there is no clear line of authority and that this could lead to managerial confusion and inefficiencies. The presence of two "masters" could be seen as a possible conflict of interest. However, field interviews indicated a general satisfaction with the arrangement from the agents, district supervisors and the NGO. Public extension agents potentially are not just passive agents of the NGO, but can bring public good types of concerns into the arrangement. Under ideal circumstances, this type of coalition represents a balance between demand-driven private sector approaches and approaches that are public good supply driven.

An additional problem, in some instances, is that Ugandan extension workers are contracted by more than one NGO and may work with the same favoured groups of farmers to carry out the activities of each NGO. The result is considerable overlap in the provision of extension services by NGOs and a concentration of services on a few farmers. A recent study supported by the Danish technical assistance agency, Danida, comparing different extension providers, including government extension, found that 24% of the farmers interviewed in Masaka district, received services from three different extension providers. In Bushenyi district, 57% of the farmers interviewed received services from more than one provider. While this may not in and of itself necessarily be wasteful duplication, it has resulted in the concentration of services on a small number of favoured farmers – only 5% of farmers in Masaka and 6% in Bushenyi were actually receiving any extension services at all.

## Concluding Comments

Despite problems, contracting-in extension coalitions reflect ongoing, field-level organizational processes. Governments are often criticised for paying for extension presence instead of for extension performance. But there is a long history of governments basically having extension services on a "retainer basis" – ensuring that they are around when needed. Other parties may "buy in" for performance, including farmers who often pay for it informally through fuel for extension workers' vehicles and food gifts. The issue may not be to try and develop organizational models for extension provision that can be imposed on local situations, but to influence and build on the "organic", ongoing organizational processes that underlie extension field activities. Decentralisation, for example, may lead to arrangements at the local level that can not be predicted a priori and from the center. As such, public–private extension coalitions merit study and analysis as part of dynamic local processes. Several areas seem to be important for closer analysis including the aspects of the local environment which foster coalitions, negotiation processes, organizational roles and how organizations perceive of each others roles, transactions costs and how to diminish them, creating collaborative platforms and managing conflicts, and measuring farm level impacts, both in terms of improved physical outcomes (e.g. profitability) and improved relationships between farmers and extension providers.

Chapter 20

# Concluding Note

William M. Rivera and Willem Zijp

Contracting for agricultural extension is a strategy employed worldwide, utilized by both public and private sector organizations. The country case studies attest to the variety of situations in which contracting has strengthened efforts to promote agricultural development and sustainability. Contracting for extension appears to be a positive development in the evolution of agricultural extension and promises to continue as a vital strategy for the advancement of knowledge transfer in the agricultural domain.

A comparison of the cases provides insight into: (i) the technical criteria involved in establishing contracts; (ii) the social and environmental issues surrounding extension contractual arrangements; and (iii) the impacts and lessons learned from contracting for extension.

## Findings on Technical Criteria

The cases reveal a number of technical criteria in contracting extension services. For instance: (i) who selects the consultant, and what criteria are used in the selection process? (ii) Who evaluates the consultant, and what criteria are used in the evaluative process? (iii) How is the total cost of the service determined; how is it shared between parties? (iv) How is quality of services controlled? (v) Who determines the contents of the extension work, and what criteria are used in the process? (vi) Who selects the endusers or "beneficiaries", and what criteria are used in this selection process? and (vii) how is the system funded; how is the budget determined?

In the *selection of advisory consultants*, individual farmers or members of farmer associations, sometimes in combination with the for-profit private sector and research institutions, are increasingly involved in the selection of the advisory consultants who serve them. Local and national governments, often in combination with non-governmental organizations (NGOs) and private companies, continue to act as primary agents in selecting advisory consultants. Local, national and international NGOs are becoming more involved, often along with governments, in selecting advisory consultants.

As for *monitoring and evaluating the advisory consultant*, government, sometimes in combination with NGOs and farmers, is generally responsible for monitoring and evaluating the work of advisory consultants. NGOs, farmers and farmer associations, are taking on greater responsibility for monitoring and evaluating the work of advisory consultants.

In most cases, government is responsible for *certifying advisory consultants*. Still, farmers and other private entities also carry out evaluations of consultants – both formally and informally.

With respect to *payment and cost sharing of the consultant service fee*, cost-sharing arrangements between governments and farmers is increasingly prevalent. In the more economically developed countries, farmers are paying either a large portion or all of the costs for advisory consultants. Farmers are engaged in cost sharing with the for-profit private sector, NGOs and research institutions, as well as with government. Sometimes the for-profit private sector and NGOs assume the costs of paying for advisory consultants.

In general, the *funding of programmes to provide contractual advisory services* is shouldered by government, sometimes in combination with farmers and the for-profit private sector. Farmers are increasingly assuming responsibility for funding contractual advisory services, and NGOs are increasingly funding programmes providing contractual advisory services to farmers. The for-profit private sector, occasionally in combination with government, also funds programmes providing contractual advisory services to farmers.

As for *who decides the content of extension messages*, farmers, in combination with NGOs, government and research institutions, are increasingly involved in making decisions as to the advisory content provided by extension consultants. Governments continue to make decisions as to the extension messages to be provided to farmers, but now more often they do so in collaboration with NGOs, and to a lesser extent with farmers and the for-profit private sector.

NGOs, as suggested above, are becoming more involved with governments, farmers and the for-profit private sector in deciding the content of extension messages.

Regarding *who decides who will receive the advisory service*, governments most often decide who will receive the advisory service, although the decision is made increasingly in conjunction with NGOs. As noted, NGOs have begun to play a larger role in decisions concerning the beneficiaries of advisory services, although they tend to do so in conjunction with government.

# Findings on Social and Environmental Issues

Social and environmental questions address large issues. The case studies address issues regarding the extent and degree of stakeholder participation, poverty, equity, food security, natural resources management, capacity building, and other concerns, such as women and children.

The capacity of the various institutions involved in the provision of advisory services through contractual arrangements was improved in most cases. Stakeholder participation in programme decision making appears to be enhanced as a result of contractual extension arrangements. Poverty issues are widely addressed in the cases. Importantly, indications are that farmer incomes increased as a result of contracts involving multiple partners. Natural resource management and environmental improvements are positive outcomes in several cases involving contractual arrangements. Integrated pest management (IPM) was noticeably the objective of extension contractual arrangements in several cases.

Contracting for extension has positively affected women farmers and labourers. Food security was enhanced in two cases, although this issue was not otherwise cited as a significant aspect of the contractual arrangements.

Equity impacts in the case studies are ambiguous. Greater equity in the distribution of benefits from the extension services was cited in three cases; however, a negative impact regarding equity of benefits for small and poor farmers was mentioned in a few cases.

# Findings on Impacts and Lessons Learned

The impacts described in the case studies reveal five domains for attention and action, including: (i) political will to promote system reorganization and contracting for extension; (ii) institutional roles, opportunities and benefits; (iii) capacity, role and utilization of providers; (iv) conditionalities; and (v) negative impacts: areas for improvement.

As in other areas of development, *political will* is required to promote system reorganization and contracting for extension. Government agencies can benefit by contracting for extension services, whether by enhancing the internal efficiency of the agency or by creating external partnerships that advance the effectiveness of advisory services. In short, government agencies need to be open to contracting with other organizations for the delivery of extension as well as with specialized consultants to provide needed services within the organization.

Government needs to spell out the *institutional roles, opportunities and benefits* in contractual arrangements (e.g. its role in training and upgrading of advisers, contract oversight, programme M&E). Those involved in contractual arrangements need to understand clearly the parameters of the reorganized system. At the same time, contracts provide an opportunity for extension providers (whether governmental, private or farmer organizations) to share experiences and resources. Collaborative planning, especially with endusers, encourages a sense of local ownership and tends to enhance programme effectiveness. Cost effectiveness and benefits to all parties involved are of course central to success. Crucial in this regard is the willingness (immediate or eventual) of endusers to pay (at least in part) for the advisory service.

Contractual arrangements tend to foster the comparative advantage of each of the different groups involved in the purposes of the advisory service while enhancing the *capacity, role and utilization of providers*. Contracts that ensure enduser participation in provider decisions about the content and delivery of extension promote demand-driven systems and the relevance of information. Training and education (through strong coalitions among agricultural education institutions, research and extension) are essential in the short term for upgrading and in the long term for advancing a professional consulting industry. Provider professionalism and technical capacity are fundamental to success. The technical competence of farmer organizations, especially their capacity to reach out to providers, is a critical factor in any extension endeavor, but particularly in those involving contractual arrangements.

Among the *conditionalities*, it is clear that contractual arrangements involve an evolutionary process and may often move through phases to reach maximum efficiency and equity (as in the cases of Chile and The Netherlands). Success depends on finding practical solutions to local particularities. A good agricultural development system (including farmer access to inputs and markets) is basic to a meaningful advisory service. Too often government services are restrained by limited funding which inhibits contractual arrangements even when the latter would be more efficient. Too often there is a lack of qualified service providers.

As for *negative impacts*, it appears that privatization and commercialization of extension services can have a negative impact on the equity of the benefits for small, poor, and marginal area farmers. There can be negative consequences for endusers not included in the contractual arrangement. Contractual arrangements can in some cases create negative competition between organizations, especially with regard to their roles in providing advisory services.

Despite its advantages and benefits, contracting should not be considered a panacea. There are disadvantages as well as advantages of utilizing non-state providers, of state sector extension providers, and of outsourcing extension services.

One disadvantage of contracting for extension services is that contractors may not understand the internal politics, the programme underpinnings, or the organizational formalities of the extension service. Also, the institution's "memory" may be forfeited through the use of contractors. Contractors are in a peripheral position to share what they have learned, and seldom influence programme strategies on a continuing basis. Also, contractors may be reluctant to pass on information about programme difficulties for fear it will reflect badly on them or limit their potential for re-hire. Costs may also prove to be more inhibitive than originally thought, especially if the need for the contractor continues beyond the time anticipated. In such a case, programme planning or management may also be at fault. Lack of continuity and institutional memory are major considerations for programme officers seeking to contract for extension services.

There are critical questions to be considered. Should in-house administrative officers and field extension staff be upgraded through special training to undertake the services required? Can the work be completed more efficiently and effectively through contracting for the service? Or, can both actions be undertaken simultaneously? If the decision is made to contract out, then in-house staff administrative officers and field extension staff should, if possible, be assigned as counterparts to consultants at comparable levels and encouraged to learn their skills.

## Legal Aspects of the Contract

When considering the legal aspects of contracts, the essentials are: (i) continuity of the extension officers involved in the contract; (ii) permanent contribution of resources by the funding source; (iii) a systematic extension methodology with a defined and validated technological package required; (iv) an efficient administrative operation with on-time allocation of payroll; (v) a rigorous system of personnel selection that ensures a well-paid staff with high technical capacity and leadership; and (vi) a continuous and tight monitoring system. Other requirements are the training of advisers – this requirement must be a precondition for certification, an initiating proactive offer of information and advice to farmers through the advisory body, and the coordination of actions for the development framework and support systems for provisions of the advisory services.

## Final Comments

Contracting for agricultural extension services reflects the advancement of public sector reform via decentralization and privatization. It also indicates the progress toward bottom-up, demand-driven, approaches to agricultural information access. Endusers are increasingly involved in extension programme decision making, e.g. in the planning, implementation and monitoring of extension programmes. The progress toward demand-driven extension tends to promote greater transparency and accountability mechanisms (e.g. certification processes, financial audits, and performance appraisals) in the administration of extension provision.

Contracting for extension services contributes to cross-sector and within-sector partnerships. The result is an increasing pluralism of extension providers. Partnerships are the result of various contemporary realities: the reduction of public sector funds and increased resource constraints in extension, diversified demands by endusers, and the overlap among organizations undertaking similar activities. Partnerships indicate the growing willingness of different sectors to work collaboratively toward the common objective of effective and efficient extension service delivery.

In some cases, contracting involves institutional pluralism where both the public and private sectors are engaged as extension providers. In Mozambique, publicly financed extension is to be opened to multiple financial and delivery arrangements, but the public sector will maintain a role in extension delivery. Thus, in some less developed countries, contracting may be an intermediate step toward involving private-sector providers, some of whom may already be operating extension systems parallel to that of the public sector.

Contracting for extension also tends to build human capacity among endusers, administrators, and extensionists (especially those who shift into advisory consultant roles). Human resources development is, along with institutional capacity building in general, a crucial factor for improving performance which in turn tends to enhance productivity, which in turn increases production.

In the 1980s, the over-extendedness of public-sector extension, the scarcities of financial resources for extension and, in some cases, lack of skilled manpower and dearth of organizational capacity (The World Bank, 1981: 5) led to major changes in ideological, economic and technical perspectives of agricultural extension. Following a period during which most government interventions for extension delivery were seen as "bad", however, critics realized that some government funding – if not government implementation – would remain necessary to ensure timely delivery of relevant advice to poor farmers who may be unable or unwilling to pay for advice that would benefit them personally or their communities. Thus, contracting for extension has come to be recognized as a logical link between public funding and private delivery.

## Benefits and Issues of Contracting for Extension Services

A number of benefits can be anticipated from contracting extension: (i) an extension service which is more operationally efficient and cost effective; (ii) a plurality of providers that can be employed to deliver extension services; (iii) extensionists who are more accountable for their performance and results; and (iv) contracted extension workers who can be rewarded for good performance and dismissed for poor performance. However, the case studies also underscore several persistent needs. These include the need for: (i) training and capacity building; (ii) active farmer involvement; (iii) clear-cut procedures and conditions for contracting extension; (iv) institutional and financial mechanisms for contracting out extension; (v) explicit conditions for grants and clarity of the qualifying criteria; and (vi) donor support to extension with greater emphasis on matching funds.

Aside from the tasks involved in organizing the process of contracting for agricultural extension, a major challenge for practitioners is that of preparing public-sector extension staff to coordinate, oversee and regulate private-sector providers and to learn from the experience of outsourcing. This new role and the accompanying responsibilities require adequate and timely external and internal support, managerial savvy at all administrative levels of public-sector extension, tight organization and decisiveness, leadership that includes participatory involvement in decision making, individual dedication and programmatic vision.

# Bibliography


Alarcón, E., Cano, J. and Moscardi, E. (eds) (1998) *Memorias del Taller sobre Situación y Perspectivas del Complejo Transferencia de Tecnologia, Asistencia Técnica y Extensión Agropecaria*. Proceedings of the Workshop of Present and Future of the Technology Transfer, Technical Assistance and Agricultural Extension. Serie Cuadernos Técnicos 3. IICA (Instituto Interamericano de Cooperación con la Agricultura). San José, Costa Rica, 300 pp.

Amanor, K. and Farrington, J. (1991) NGOs and agricultural technology development. In: Rivera, W.M. and Gustafson, D.J. (eds) *Agricultural Extension: Worldwide Institutional Evolution and Forces for Change*. Elsevier, Amsterdam.

Amaro, P. (1999) *Para a Optimização da Protecção Integrada e da Produção Integrada até 2006*. Universidade Técnica de Lisboa, Instituto Superior de Agronomia, Lisboa.

Ameur, C. (1994) *Agricultural Extension: a Step Beyond the Next Step*. World Bank Technical Paper no. 247. The World Bank, Washington, DC.

Anderson, J. (1999) *Extension, Education, Training and Research*. Backstopping mission to Mozambique, Community Forestry and Wildlife Management GCP/MOZ/056/NET. FAO, Rome.

Anderson, J., Crowder, V. and Clement, J. (1999) Pluralism in sustainable forestry and rural development – an overview of concepts, approaches and future steps. In: *Pluralism and Sustainable Forestry and Rural Development*. FAO, Rome, pp. 1-56.

Antholt, C.H. (1994) *Getting Ready for the Twenty First Century: Technical Change and Institutional Modernization in Agriculture*. World Bank Technical Paper no. 217. The World Bank, Washington, DC.

Antholt, C. (1994). *Modernization of Agricultural Extension in Turkey: Strategic Issues, Commercialization and Privatization*. The World Bank, Washington, DC.

Antholt, C. and Zijp, W. (1995) Keys to the success of farmer participatory involvement in extension. In: *The World Bank Participation Sourcebook*. The World Bank, Washington, DC.

Ashworth, V.A. (1999) *Extension and the Private Sector: Expanding the Private Sector Role as Extension Providers: Principles, Issues, Options and Actions*. World Bank, Government of Zambia, Lusaka.

Ashworth, V.A. (1999) *Extension at the Crossroads: a Review of Agricultural Extension in Zambia*. World Bank, Government of Zambia, Lusaka.

ASIRP (1999) Project appraisal document. Agricultural Services Innovation and Reform Project (ASIRP), World Bank report no. 19662-BD.

Avenriep, G. (1997) Einführung von Beratungsgebühren in Westfalen-Lippe. *Ausbildung und Beratung* 4, 82-84.

Bangladesh Department of Agricultural Extension (DAE) (1999) *Strategic Plan, 1999-2002*. Khamarbari, Dhaka, Bangladesh.

Bastos, M. (1997) *La Asistencia Técnica: Una Nueva Panacea para el Desarrollo Agroalimentario y Rural?* Presented at Taller: Situación Actual y Perspectivas del Complejo Transferencia de Technología, Asistencia Técnica y Extensión Agropecuaria, San Jose, Costa Rica.

Bathrick, D.D. (1997) *Fostering Global Well-being: a New Paradigm to Revitalize Agricultural and Rural Development*. Food, Agriculture, and the Environment discussion paper no. 26. IFPRI, Washington, DC.

Bebbington, A. (1989) *Institutional Options and Multiple Sources of Agricultural Innovation: Evidence from an Ecuadorian Case Study*. ODI Agricultural Administration (Research and Extension) Network paper no. 11, London.

Bell, A. (1998) Changing from adviser to consultant: the consultant's experience. In: Wallace, I. (ed.) *Rural Knowledge Systems for the 21st Century: Rural Research and Extension in Western, Central and Eastern Europe*. AERDD, Reading, UK, pp. 199-203.

Bennett, C.F. (1996) Rationale for public funding of agriculture extension programs. *Journal of Agricultural and Food Information* 3(4), 3-25.

Bentley, J.W. and Baker, P.S. (2000) *The Colombian Coffee Growers' Federation: Organised, Successful Smallholder Farmers for 70 Years*. Network Paper no. 100. Agricultural Research and Extension Network, Overseas Development Institute, London.

Beynon, J. *et al.* (1998) *Financing the Future: Options for Research and Extension in Sub-Saharan Africa*. Oxford Policy Management, Oxford, UK.

Boehje, M. (1998) Information and technology transfer in agriculture: the role of the public and private sectors. In: Wolf, S.A. (ed.) *Privatization of Information and Agricultural Industrialization*. CRC Press, Boca Raton, Florida.

Bos, J.T.M. (1989) Privatization and reorganization of the Dutch agricultural extension service. In: *Agrarische Voorlichting*, no. 5. Ministry of Agriculture and Fisheries, The Hague.

Bos, J.T.M., Proost, M.D.C. and Kuiper, D. (1991) Reorganizing the Dutch Agricultural Extension Service: the IKC in focus. In: Kuiper, D. and Röling, N.G. (eds) *Proceedings of the European Seminar on Knowledge Management and Information Technology*. Wageninen Agricultural University, Department of Extension Service, Wageninen, The Netherlands, pp. 65-73.

Bunney, F.M.G. and Bawcutt, D.E. (1991) Making a business of an extension service. *Agricultural Progress* 66, 36-43.

Buttel, F. (1991) The restructuring of the American public agricultural research and technology transfer system: implications for agricultural extension. In: Rivera, W.M. and Gustafson, D.J. (eds) *Agricultural Extension: Worldwide Institutional Evolution and Forces for Change*. Elsevier Science Publishers, Amsterdam.

Carney, D. (1998) *Changing Private and Public Roles in Agricultural Service Provision*. Overseas Development Institute, Natural Resources Group, London.

Carney, D. and Farrington, J. (1998) *Natural Resource Management and Institutional Change*. Routledge, London.

Cary, J.W. (1993) Three eras of extension in Australia: from public folly to private good. In: *Australia-Pacific Extension Conference Proceedings*, Vol. 1. Department of Primary Industries, Brisbane, Australia.

Chabala, C. (1998) *Peoples Participation Programme in Western Zambia: Successful Lessons and Innovative Experience*. Food and Agriculture Organization, Lusaka.

Christoplos, I. (1995) *Paradigms, Policy and Privatisation in Vietnamese Agricultural Extension*. Working paper no. 275. Swedish University of Agricultural Sciences.

Christoplos, I. and Nitsch, U. (1996) *Pluralism and the Extension Agent: Changing Concepts and Approaches in Rural Extension*. Swedish University of Agricultural Sciences.

Ciba Foundation Symposium (1997) *Precision Agriculture: Spatial and Temporal Variability of Environmental Quality*. John Wiley & Sons, New York.

Clarke, T. (1994) Reconstructing the public sector: performance measurement, quality assurance, and social accountability. In: Clarke, T. (ed.) *International Privatisation: Strategies and Practices*. Walter de Gruyter, Berlin/New York.

Clarke, T. (ed.) (1994) *International Privatisation: Strategies and Practices*. Walter de Gruyter, Berlin/New York.

CNRA (1999) Atelier sous-regional sur le partenariat recherche agricole. Rapport de synthese, CNRA, Organisations de Producteurs, Vulgarisation Agricole, 18 pp.

CORAF/ODI/CIRAD/ITAD (1999) Renforcer la collaboration entre la vulgarisation et les organisations paysannes en Afrique de l'Ouest et du centre. Synthese des propositions, CORAF, 50 pp. + Annexes.

CORAF/ODI/CIRAD/ITAD (1999) Renforcer la collaboration entre la vulgarisation et les organisations paysannes en Afrique de l'Ouest et du centre. Etude de terrain - Burkina Faso, CORAF, 62 pp. + Annexes.

CORAF/ODI/CIRAD/ITAD (1999) Renforcer la collaboration entre la vulgarisation et les organisations paysannes en Afrique de l'Ouest et du centre. Etude de terrain - Cameroun, CORAF, 57 pp. + Annexes.

CORAF/ODI/CIRAD/ITAD (1999) Renforcer la collaboration entre la vulgarisation et les organisations paysannes en Afrique de l'Ouest et du centre. Etude de terrain - Guinee, CORAF, 53 pp. + Annexes.

Coulter, J., Goodland, A., Tallontire, A. and Stringfellow, R. (1999) *Marrying Farmer Co-operation and Contract Farming for Service Provision in a Liberalising Sub-Saharan Africa*. Natural Resource Perspectives, 48. Overseas Development Institute, London.

Crowder, L. Van (1987) Agents, vendors and farmers: public and private sector extension in agricultural development. *Agriculture and Human Values* Fall, 26.

Crowder, L. Van (1991) Extension for profit: agents and sharecropping in the highlands of Ecuador. *Human Organization* 50(1).

Crowder, L. Van (1998) World Bank Agricultural Extension Project II – Design and Preparation Mission, Uganda. FAO, Rome.

Crowder, L. Van (1999) Uganda Agricultural Extension Project Preparation Mission. FAO, Rome.

Dillinger, B. (1995) Decentralization, politics and public services. In: Estache, A. (ed.) *Decentralizing Infrastructure: Advantages and Limitations*. The World Bank, Washington, DC.

Donahue, J.D. (1989) *The Privatization Decision: Public Ends, Private Means*. Basic Books, New York.

Duijsings, P. (1998) Information and communication management in a privatised extension organisation: the case of the DLV. In: *Assessing the Impact of Information and Communication Management on Institutional Performance*. Proceedings of a CTA workshop, Wageningen, The Netherlands, pp. 59-68.

Edmonds, C.M. (1998) Policy regimes, agrarian institutions, and the performance of smallholder agriculture in Chile: three essays analyzing longitudinal survey data on Chilean peasant farms. Thesis. University of California, Berkeley.

Eicher, C.K. (1999) *Institutions and the African Farmer*. Issues in Agriculture 14. Consultative Group on International Agricultural Research, Washington, DC.

Farrelly, M. (1996). *A Study of Participation in Chivi Food Security Project*. Intermediate Technology Development Group (ITDG), Harare, Zimbabwe.

Farrington, J. (1994) *Public Sector Agricultural Extension: is Their Life After Structural Adjustment? Natural Resources Perspectives*, no. 2, Nov. Overseas Development Institute, London.

Farrington, J. (1998) *Organisational Roles in Farmer Participatory Research and Extension: Lessons from the Last Decade. Natural Resource Perspectives,* no. 27. Overseas Development Institute, London.

Feder, G., Willet, A. and Zijp, W. (1999) *Agricultural Extension: Generic Changes and Some Ingredients for Solutions*. Policy Research Working paper no. 2129. The World Bank, Washington, DC.

Food and Agriculture Organization (FAO) (1989) *Report of the Global Consultation on Agricultural Extension*. FAO, Rome.

Freeman, O. (1989) Reaping the benefits: cash crops in the development process. *Health and Development* 1(1), 21-23.

Gautam, M. and Anderson, J.R. (1999*) Reconsidering the Evidence on the Returns to T&V Extension in Kenya*. Operations Evaluations Department, The World Bank, Washington, DC.

Goe, W.R. and Kenney, M. (1988) The political economy of the privatization of agricultural information: the case of the United States. *Agricultural Administration and Extension* 28, 81-99.

Gómez, S. (1991) Nuevas modalidades de apoyo a la pequeña agricultura. El caso de Chile. *Estudios Sociales* 70, 131-147.

Grygo, H. (1996) 'Landwirtschaftliche Beratung - quo vadis?' *Ausbildung und Beratung* 5(96), 88-89.

Gui, B. (1994) On the third sector in Central and Eastern European post-Soviet type economies. In: Clarke, T. (ed.) *International Privatisation: Strategies and Practices*. Walter de Gruyter, Berlin/New York, pp. 273-288.

Hercus, J.M. (1991) The commercialization of government agricultural extension services in New Zealand. In: Rivera, W.M. and Gustafson, D.J. (eds) *Agricultural Extension: Worldwide Institutional Evolution and Forces for Change*. Elsevier Science Publishers, Amsterdam.

Hoffmann, V. (1995, 1996) Landwirtschaftliche beratung: wohin. *Ausbildung und Beratung*. Part 1; 12(95), 227-229. Part 2; 1(96), 10-12.

Hoffmann, V., Lamers, J. and Kidd, A.D. (2000) *Reforming the Organization of Agricultural Extension in Germany: Lessons for Other Countries*. AgREN Network Paper no. 98. Overseas Development Institute, London.

Hommes, R. (1995) Conflicts and dilemmas of decentralization. Paper presented at the Annual World Bank Conference on Development Economics. The World Bank, Washington, DC.

Hortex (2000) Horticultural Export Development Foundation (Hortex): an introduction. Unpublished. House no. 3/1, Block -B, Lalmatia, Dhaka, Bangladesh.

Howard, J., Kelly, V., Maredia, M., Stepanek, J. and Crawford, E. (1999) *Progress and Problems in Promoting High External Input Technologies in Sub-Saharan Africa: the Sasakawa Global 2000 Experience in Ethiopia and Mozambique.* Staff Paper no. 99-24. Michigan State University, Department of Agricultural Economics, East Lansing, Michigan.

Howard, J., Jeje, J.J., Tschirley, D., Strasberg, P., Crawford, E. and Weber, M. (1998) *Is Agricultural Intensification Profitable for Mozambican Smallholders? An Appraisal of the Inputs Subsector and the 1996/97 DNER/SG2000 Program* Research Report no. 31. Ministry of Agriculture and Fisheries, Directorate of Economics, Maputo.

Howell, J. (1985) *Recurrent Costs and Agricultural Development.* Overseas Development Institute, London.

Huang, R.Q. (1992) The level of cooperation among agricultural extension organizations in the greenhouse vegetable sector in the westland area in The Netherlands. Paper presented at the 8th World Congress for Rural Sociology, 11-16 August. Pennsylvania State University, University Park, Pennsylvania.

INCRA (1997) *Manual of LUMIAR* (In Portuguese). INCRA, Brasilia, DF.

INDAP (1992) *Improving the Transfer of Technology Program* (In Spanish). Instituto de Desarrollo Agropecuario, Santiago, Chile, 25 pp.

Intermediate Technology Development Group (ITDG) (1997) *Chivi Food Security Project: a Process Approach.* ITDG, Lusaka.

Jiggins, J. (1989) International Seminar on Rural Extension Policies. Seminar proceedings, Part 1, 26-30 June. IAC, Washington, DC.

Johansen, R. and Swigart, R. (1994) *Upsizing the Individual in the Downsized Organization.* Addison-Wesley, Reading, Massachusetts.

Kalaitzandonakes, N. and Bullock, J.B. (1998) Technology and information transfer in U.S. agriculture: the role of land grant universities. In: Wolf, S.A. (ed.) *Privatization of Information and Agricultural Industrialization.* CRC press, Boca Raton, Florida.

Kamerman, S.B. and Kahn, A. (eds) (1994) *Privatization and the Welfare State.* Princeton University Press, Princeton, New Jersey.

Katz, E. (1999) The three "C": conditions for functioning voucher systems in agricultural extension. *BeraterInnen News* 2(99), 13-16.

Kloppenburg, J., Jr (1988) *First the Seed.* Cambridge University Press, New York.

Kraft, N.J. (1997) Public and private extension: who pays and who provides? Paper presented at the annual conference of the South African Society for African Extension, Adventura Warmbaths, South Africa.

Lanz, M. (1993) Gruppenberatungsansätze im Agrarsektor. *Berichte über Landwirtschaft* 71, 98-105.

Le Gouis, M. (1991) Alternative financing of agricultural extension: recent trends and implications for the future. In: Rivera, W.M. and Gustafson, D.J. (eds) *Agricultural Extension: Worldwide Institutional Evolution and Forces for Change.* Elsevier Science Publishers, Amsterdam.

Leonard, D.K. (1985) African Practice and the Theory of User Fees. In: Howell, J. (ed.) *Recurrent Costs and Agricultural Development.* Overseas Development Institute, London, pp. 130-145.

Li, X., Liu, Y. and Li, O. (1996) Development tendency of Chinese agricultural extension. *Agricultural Modernization* 3.

Liu, Y. (1998) *Institutional and Policy Reform of Rural Extension in China During the Transition towards a Market Economy. Training for Agriculture and Rural Development (TARD).* FAO, Rome.

Maalouf, W.D., Adhikarya, R. and Contado, T. (1991) Extension coverage and resource problems: the need for public-private cooperation. In: Rivera, W.M. and Gustafson, D.J. (eds) *Agricultural Extension: Worldwide Institutional Evolution and Forces for Change.* Elsevier Science Publishers, Amsterdam.

Madon, G. (1996) Gestion locale sécurisée des ressources renouvelables et du foncier (GELOSE). Etude de faisabilité. Office national de l'environnement, Madagascar.

Mahdavi, G. (1985) A commodity-driven approach: the experience of the Compagnie Française pour le Développement de Textiles - CFDT. Adapted from papers presented at the African Workshop on Extension and Research, Eldoret, Kenya, 10-16 June 1984, and at the West Africa Seminar on Agricultural Extension and its Link with Agricultural Research in Rural Development, Yamoussoukro, Côte d'Ivoire. CFDT, Paris.

Martin, J. (1995) Can government be reinvented? In: Boston, J. (ed.) *The State Under Contract.* Bridget Williams Books Ltd, Wellington, New Zealand.

Mathews, J.T. (1997) Power shift. *Foreign Affairs* 76(1), 50-66.

Mercoiret, M.R. (1994) *L'appui aux Producteurs Ruraux.* Karthala, Paris.

Monardes, A., Escobar, G. and González, G. (eds) (1994) *Transferencia de Tecnología Agropecuaria. De la Generación de Recomendaciones a la Adopción: Enfoques y Casos.* (Transfer of agricultural technology: from the generation of recommendations to adoption: approaches and case studies.) IDRC, RIMISP, Santiago, Chile, 207 pp.

Moris, J. (1991) *Extension Alternatives in Tropical Africa.* Occasional paper no. 7. ODI, London.

Mosher, A.T. (1979) *Introduction to Agricultural Extension.* Agricultural Development Council, New York (available from Winrock International, Morrillton, Arkansas).

Namdar-Iraní, M. and Quezada, X. (1995) Formulación de propuestas locales de desarrollo silvoagropecuario para los principales sistemas de producción campesinos chilenos. In: Berdegué, J.A. and Ramírez, E. (eds) *Investigación con Enfoque de Sistemas en la Agricultura y el Desarrollo Rural.* RIMISP, Santiago, Chile, pp. 147-158.

National Research Council (1997) *Precision Technology in the 21st Century.* National Academy Press, Washington, DC.

Neilson, D. (1992) Institutional reform and privatization in research and extension. Paper presented at Latin American Agricultural Retreat (Harpers Ferry, West Virginia). The World Bank, Washington, DC.

Neuchatel Group (1999) *Analyse de Quelques Approaches de Vulgarisation Agricole en Afrique au Sud du Sahara.* Direction Générale de la Coopération Internationale et du Développement, Bureau des Politiques Agricôles et de la Sécurité Alimentaire, Paris.

Neuchatel Group (1999) *Common Framework for Agricultural Extension.* Direction Générale de la Coopération Internationale et du Développement, Bureau des Politiques Agricôles et de la Sécurité Alimentaire, Paris.

Odgaard, J.D. (1999) Agricultural extension service in Uganda – a comparison of three extension systems in Masaka and Bushenyi Districts. MSc thesis. The Royal Veterinary and Agricultural University, Denmark.

OECD (Organization of Economic Cooperation and Development) (1989) *Survey on Effects and Consequences of Different Forms of Funding Agricultural Services.* OECD doc. AGR/REE 89, 7. OECD, Paris.

Parker, A.N. (1995) *Decentralization: the Way Forward for Rural Development.* Agricultural and Natural Resources Department, The World Bank, Washington, DC.

Paul, B. (1997) *Mission Report: Training and Extension Consultancy for UNDP/FAO Project 96/014 Building Capacity for Smallholder Forestry Development.* Hanoi, Vietnam.

Pray, C. and Echeverría, R. (1990) Private sector agricultural research and technology transfer links in developing countries. In: Kaimowitz, D. (ed.) *Making the Link: Agricultural Research and Technology Transfer in Developing Countries.* Westview Press, Boulder, pp. 197-226.

PROAGRI (1998) *Mozambique: PROAGRI Appraisal Mission* (Extension Component). FAO, Rome.

PROCASUR, GTZ, FIDA, GIA, AECI y PRODEPA (1997) *Seminario Latinoamericano: Sistemas Privados de Asistencia Técnica.* Santa Cruz de la Sierra, Bolivia.

Proost, J. and Matteson, P. (1997) Integrated farming in the Netherlands: flirtation or solid change? *Outlook on Agriculture* 26(2), 87-94.

Proost, J. and Röling, N. (1991) 'Going Dutch' in extension. *Interpaks Interchange* 9(1), 3-4.

Ramahazomanana, S. (1996) Processus d'intégration de la population riveraine dans la gestion des forêts d'état - station forestière de Manjakatompo. Projet Germano-Malgache de Développement Forestier Intégré dans la Région de Vakinankaratra (PDFIV), Madagascar.

Randrianasolo, J. (1996) Etude de faisabilité d'un projet d'appui à la gestion participative de forêts villagoises dans la région de Port Bergé. Madagascar.

Rapp, L. (1986) *Techniques de Privatisation des Entreprises Publiques.* Librairies Techniques, Paris.

Rivera, W.M. (1991) Worldwide policy trends in agricultural extension (Parts I and II). *The Journal of Technology Transfer* 16(1), 13-18; 16(3), 29-37.

Rivera, W.M. (1996) Reinventing agricultural extension: fiscal system reform, decentralization, and privatization. *Journal of International Agricultural and Extension Education* 3(1), 63-67.

Rivera, W.M. (1996) Agricultural extension in transition worldwide: structural, financial and managerial strategies for improving agricultural extension. *Public Administration and Development* 16, 151-161.

Rivera, W.M. (1996) Agricultural extension into the next decade. *European Journal of Agricultural Education and Extension* 2(4).

Rivera, W.M. (1997) Confronting the global market: public sector agricultural extension reconsidered. In: *The Current Situation in and the Outlook for the Technology Transfer, Technical Assistance and Agricultural Extension Complex Proceedings.* IICA, San José, Costa Rica.

Rivera, W.M. and Cary, J.W. (1998) Privatising agricultural extension worldwide: institutional changes in funding and delivering agricultural extension. In: Swanson, B.E. (ed.) *Agricultural Extension:Reference Mannual,* 3rd edn. FAO, Rome.

Rivera, W.M. and Gustafson, D.J. (eds) (1991) *Agricultural Extension: Worldwide Institutional Evolution and Forces for Change.* Elsevier, Amsterdam, The Netherlands.

Rogers, W.L. (1987) The private sector: its extension systems and public/private coordination. In: Rivera, W.M. and Schram, S.G. (eds) *Agricultural Extension Worldwide: Issues, Practices and Emerging Priorities.* Croom Helm, New York.

Röling, N. (1988) *Extension in Europe and the Third World: Comparisons and Implications.* Occasional Papers in Rural Extension no. 3. University of Guelph, Department of Rural Extension Studies, Guelph, Canada.

Röling, N. (1996) Towards an interactive agricultural science. *European Journal of Agricultural Education and Extension* 2(4).

Scheuermeier, U., Zellweger, T. and Schmidt, P. (1998) Enhancing accountability by reversing the flow of funds: results of two think-tanking workshops illustrated with examples from the field. *AgREN Newsletter* 38.

Schwartz, L.A. (1994) *The Role of the Private Sector in Agricultural Extension: Economic Analysis and Case Studies.* Agricultural Administration (Research and Extension) Network paper no. 18. Overseas Development Institute, London.

Scott, G.C. (1996) *Government Reform in New Zealand.* Occasional paper no. 140. International Monetary Fund, Washington, DC.

Selener, D., Chenier, J. and Zelaya, R. (1997) *Farmer to Farmer Extension: Lessons from the Field* (English translation). International Institute for Rural Reconstruction, Quito.

Sepúlveda, S. (ed.) (1995) *Memorias del Taller Sobre Transferencia de Tecnología Apropiada Para Pequeños Productores con Métodos Participativos* (Proceedings of the Workshop on Transfer of Appropriate Technology for Small Farmers with Participatory Methods). CIID/GTZ/IICA, San José, Costa Rica, 380 pp.

Smith, L.D. (1997) *Decentralisation and Rural Development.* FAO/SARD, Rome.

Stringfellow, R. and McKone, C. (1996) *The Provision of Agricultural Services through Self-help in Sub-Saharan Africa: Zimbabwe Case Study.* Overseas Development Administration, Natural Resources Institute, Plunket Foundation, London.

Tacken, W.A.G. (1991) Why and how did we change our Agricultural Advisory Service approach in The Netherlands? In: *Challenges and Threats for Agricultural Advisory Services Created by Global Market Situation and New Emphasis on Rural Areas.* University of Helsinki, Helsinki, Finland, pp. 46-57.

Tacken, W.A.G. (1996) Landbouwvoorlichting, wie betaalt? In: *Proceedings of the International Agriculture Day 'De Prijs van Landbouwvoorlichting'.* Royal Tropical Institute Amsterdam and Royal Agricultural Association, Wageningen, The Netherlands, pp. 23-32.

Tendler, J. (1997) *Good Government in the Tropics.* The Johns Hopkins Press, Baltimore, Maryland.

Thomson, A.M. and Smith, D. (1990) Privatization of distribution of inputs, services and marketing of agricultural output. A paper prepared for the Human Resources, Institutions and Agrarian Reform Division, FAO, Rome.

Thompson, E., Jr (1986) *Small is Bountiful.* American Farmland Trust, Washington, DC.

Uganda, Government of (1998) *Plan for Modernization of Agriculture.* Kampala.

Umali-Deininger, D. (1996) *The Public and Private Sector in Agricultural Extension: Partners or Rivals?* The World Bank, Washington, DC.

Umali-Deininger, L. and Schwartz, L. (1994) *Public and Private Agricultural Extension: Beyond Traditional Frontiers.* Technical paper no. 247. The World Bank, Washington, DC.

USAID, ONE, DEF, TRandD, ANGAP (1996) Ateliers de discussion de la mise en oeuvre de la gestion locale sécurisée (GELOSE) des ressources naturelles renouvelables. Hôtel Résidence, Madagascar.

US Department of Agriculture. Extension Committee on Organisation and Policy (ECOP). Futures Task Force (1987) *Extension in Transition: Bridging the Gap between Vision and Reality.* Washington, DC.

US Office of Technology Assessment (OTA) (1986) *Technology, Public Policy and the Changing Structure of American Agriculture.* Washington, DC.

Van den Ban, A. (2000) *Different Ways of Financing Agricultural Extension.* Agricultural Research and Extension Newsletter. Overseas Development Institute, London.

Von Braun, J. (1993) *Urban Food Security and Malnutrition in Developing Countries: Trends, Policies and Research Implications.* Occasional Paper no. 27. IFPRI, Washington, DC.

Walker, A.B. (1993) *Recent New Zealand Experience in Agricultural Extension.* Proceedings of the Australia-Pacific Extension Conference, Gold Coast, Australia.

White, R. and Eicher, C.K. (1999) *NGO's and the African Farmer: a Sceptical Perspective.* Staff Paper no. 99-01. Michigan State University, Department of Agricultural Economics, East Lansing, Michigan.

Whiteside, M. (1998) *Encouraging Sustainable Smallholder Agriculture in Southern Africa in the Context of Agricultural Services Reform.* Natural Resource Perspectives no. 36. Overseas Development Institute, London.

Wilson, M. (1991) Reducing the costs of public extension services: initiatives in Latin America. In: Rivera, W.M. and Gustafson, D.J. (eds) *Agricultural Extension: Worldwide Institutional Evolution and Forces for Change.* Elsevier Science Publisher, Amsterdam.

Wise, C.R. (1990) Public service configurations and public organizations: public organization design in the post-privatization era. *Public Administration Review* 50(2), 141-155.

Wolf, S.A. (ed.) (1998) *The Privatization of Information and the Industrialization of Agriculture.* CRC Press, Boca Raton, Florida.

World Bank (1990) *Agricultural Extension: the Next Step.* Policy and Research Series, Washington, DC.

World Bank (1991) *Developing the Private Sector: the World Bank's Experience and Approach.* Country Economics Department, Washington, DC, pp. 12-13.

World Bank (1994) *Chile: Strategy for Rural Areas. Enhancing Agricultural Competitiveness and Alleviating Rural Poverty.* Report no. 12776-CH. Natural Resources and Rural Poverty Division, Washington, DC.

World Bank (1995) *Participation Sourcebook.* Washington, DC, pp. 208-211.

Zijp, W. (1991) *From Extension to Agricultural Information Management: Issues and Recommendations from World Bank Experience in the Middle East and North Africa.* The World Bank, Agriculture Division, Technical Department, Washington, DC.

Zijp, W. (1994) *Improving the Transfer and Use of Agricultural Information: a Guide to Information Technology.* Discussion Paper. The World Bank, Washington, DC.

# Web Site References

## Development Organizations and Research

http://www.idrc.ca/library/world/world.html
Directory on development produced by the International Development and Research
  Council of Canada.

http://ntr.ids.ac.uk/eldis
Information on development and the environment, supported by Danish Aid and hosted
  by the Institute of Development Studies at Sussex University, England.

http://www.id21.org
Information about UK-based or UK-funded research on development issues.

http://www.oneworld.org
"Membership" site which contains the web pages of many development organizations
  from around the world, including most UK NGOs.

http://www.worldbank.org/
Direct link to the World Bank homepage.

http://www.fao.org/
Direct link to the Food and Agriculture Organization/UN.

http://www.ifad.org/
Direct link to the International Fund for Agricultural Development.

http://www.iica.ac.cr
Direct link to the Inter-American Institute for Cooperation on Agriculture.

http://www.cgiar.org/
Direct link to the Consultative Group on International Agricultural  Research and its 16
    International Agricultural Research Centers (IARCs).

http://www.usaid.org
Direct link to the US Agency for International Development.


## Information on the Various Continents

http://www.PanAsia.org.sg/index2.html

http://lanic.utexas.edu
Links to Latin America

http://www.sul.stanford.edu/depts/ssrg/africa/guide.html
Links to Africa


## Index to Newspapers Available On-line from Across the World

http://excite.netscape.com/directory/news/world_newspapers

# Index